Advances in Spectroscopy and Imaging of Surfaces and Nanostructures

MATERIALS RESEARCH SOCIETY
SYMPOSIUM PROCEEDINGS VOLUME 1318

Advances in Spectroscopy and Imaging of Surfaces and Nanostructures

Symposium held November 29–December 3, Boston, Massachusetts, U.S.A.

EDITORS

John Cumings
University of Maryland
College Park, Maryland, U.S.A.

Dillon D. Fong
Argonne National Laboratory
Lemont, Illinois, U.S.A.

Jianyu Huang
Sandia National Laboratories
Albuquerque, New Mexico, U.S.A.

Stuart Lindsay
Arizona State University
Tempe, Arizona, U.S.A.

Guangwen Zhou
State University of New York, Binghamton
Binghamton, New York, U.S.A.

Jinghua Guo
Lawrence Berkeley National Laboratory
Berkeley, California, U.S.A.

Hendrik Bluhm
Lawrence Berkeley National Laboratory
Berkeley, California, U.S.A.

Michael Hävecker
Fritz Haber Institute of the
Max Planck Society
Berlin, Germany

Shu Yamaguchi
University of Tokyo
Tokyo, Japan

Gertjan Koster
University of Twente
Enschede, The Netherlands

Fabio Miletto Granozio
dell' Università di Napoli Federico II
Napoli, Italy

Gyula Eres
Oak Ridge National Laboratory
Oak Ridge, Tennessee, U.S.A.

Chang-Beom Eom
University of Wisconsin–Madison
Madison, Wisconsin, U.S.A.

Nicholas Ingle
University of British Columbia
Vancouver, Canada

Oleg V. Kolosov
Lancaster University
Lancaster, United Kingdom

Bryan D. Huey
University of Connecticut
Storrs, Connecticut, U.S.A.

Seungbum Hong
Argonne National Laboratory
Lemont, Illinois, U.S.A.

Hyunjung Shin
Kookmin University
Seoul, South Korea

MRS
Materials Research Society
Warrendale, Pennsylvania

CAMBRIDGE
UNIVERSITY PRESS

CAMBRIDGE UNIVERSITY PRESS
Cambridge, New York, Melbourne, Madrid, Cape Town,
Singapore, São Paulo, Delhi, Tokyo, Mexico City

Cambridge University Press
32 Avenue of the Americas, New York, NY 10013-2473, USA

www.cambridge.org
Information on this title: www.cambridge.org/9781605112954

Materials Research Society
506 Keystone Drive, Warrendale, PA 15086, USA
http://www.mrs.org

First published 2011

Financial support for the Symposium TT was provided by:
SPECS Surface Nano Analysis GmbH, Voltastrasse 5, 13355 Berlin, Germany
InSync, Inc., 2511C Broadbent Pkwy., Albuquerque, NM 87107

CODEN: MRSPDH

ISBN: 978-1-60511-295-4 Hardback

CONTENTS

*Invited Paper

IN SITU X-RAY SYNCHROTRON RADIATION SPECTROSCOPIES IN ENERGY-RELATED MATERIALS SCIENCE AND HETEROGENEOUS CATALYSIS

REAL-TIME STUDIES OF EVOLVING THIN FILMS AND INTERFACES

*Invited Paper

NOVEL DEVELOPMENT AND APPLICATIONS OF SCANNING PROBE MICROSCOPY

*Invited Paper

PREFACE

There are, perhaps, three components that are vital to the success of any scientific quest. They are 1) creative thinking and courageous experimentation of researchers in the field, 2) an amalgamation of this research with approaches of the broader research community (which often creates the most exciting ideas), and, finally, 3) practical applications of the research that put it in the broader perspective of our lives.

All of these components are distinctive features of four symposia that were held at the Fall MRS 2010 meeting in Boston and are brought together in this volume, namely, Symposium SS: "Advanced Imaging and Scattering Techniques for In Situ Studies", TT: "In Situ X-Ray Synchrotron Radiation Spectroscopies in Energy-Related Materials Science and Heterogeneous Catalysis", UU: "Real-Time Studies of Evolving Thin Films and Interfaces" and VV: "Novel Development and Applications of Scanning Probe Microscopy." A major unifying theme for these symposia is exploration of intricate properties of materials on the near-to-atomic length scale in the immediate vicinity of the free surface or at interfaces between materials. These symposia focus on various aspects and approaches of exploration of surfaces and interfaces from more traditional electron and x-ray scattering (the focus of symposia SS, TT, and UU) to more recent and rapidly advancing scanning probe microscopy, or SPM (Symposium VV). Together, they cover a field of great importance across the broad MRS community.

Of tremendous interest in modern material science, this key aspect of nanotechnology considers how interfaces and surfaces help to determine properties and functionalities of a wide range of materials, including oxides, metals and nanoparticles. Monitoring evolving surfaces in real time is a prerequisite for mastering the evolution process itself. Furthermore, surfaces and interfaces often display surprising properties that are qualitatively different from those of the corresponding bulk materials. Controlling and tailoring the physical properties at the interfaces between different materials on the atomic scale can therefore result in real scientific breakthroughs. Research in atomic control when fabricating interfaces and surfaces will help in understanding important structure-property relationships as well as in improving the design of nano-devices. Another aspect of this research is developing techniques to probe the structure and/or the properties of evolving surfaces and interfaces in environments where they are created and used.

Symposium UU was a forum for researchers who use or develop in-situ characterization and monitoring techniques for thin films, surfaces and interfaces. Experts in the fields of real-time 1) scattering (e.g., XRD and TEM), 2) imaging (e.g., SPM) and 3) spectroscopy (e.g., XPS) came together to report on their respective progress and/or new developments. Different kinds of probes, such as light in a wide energy spectrum, electrons and scanning probes, and their various capabilities to adapt, case by case, to the very peculiar working conditions required for this research emerged. Also, a particular opportunity was given for researchers to show prospects for applications of novel techniques to real-time analysis.

The search for new sources of clean energy is rapidly becoming one of the most pressing technological challenges that we face today, and this was the focus of Symposium TT. Enormous progress has been made in developing new, tailored materials via nanostructuring, self-assembly and bio-mimetic methods, that are key for developing renewable energy sources, such as solar energy conversion, as well as for the rational design of highly efficient catalysts. The effort requires new materials developed through the control of the atomic, chemical and electronic structure, which in turn requires an intimate collaboration between materials synthesis and characterization of the electronic properties of complex materials. The presentations at this symposium show the opportunities that synchrotron radiation research can provide to answer some of the challenges associated with energy science research, including exploration of high temperature superconductors, lithium ion batteries, and fuel cells, to name a few. These materials come in a broad range of chemical and structural complexity, rendering their analysis and diagnostics difficult. A large number of critical issues relevant for energy purposes have been addressed with x-ray scattering and spectroscopy including devices in operation.

Finally, Symposium VV focused on novel SPM-related advances that are a vital and, often the only, tool allowing precise imaging and characterization of certain nano-structures. The papers spanned broad areas from chemical identification at atomic resolution (Mohn, Pethica, Schwarz) and new probe functionalities (Ko, Westervelt, Ashby, Hong) to nanomechanical (Carpick, Hurley, Kolosov), thermal (Narayanaswamy, King), optical (Wickramasingh, Haugstad) and electrical (Bonnell, Kim, Kalinin, Shin) SPM measurements. Novel developments in nanomanufacturing (Weaver, King, Li) and new high speed and dynamic measurements (Besenbacher, Miles, Frenken, Huey) showed their increasing impact on the future applications of SPM. The panel discussion chaired by Dawn Bonnell involved lead researchers of major scanning probe instrumentation companies, government agencies and leading researchers and touched upon the shortcomings, challenges and inspiration of this relatively young developing field.

The distinctive offering of Symposium VV was a panel of developers and users of SPM, also chaired by Dawn Bonnell. Given the close link between all four symposia, we held several joint sessions, allowing deep integration of research in this valuable area of material science. Overall, our symposia brought together the leading-edge expertise of various characterization methods, nanoscale and advanced materials manufacturing and fundamental electronic and chemical properties of the materials involved with a great selection of invited and contributed talks, eight large poster sessions and a panel discussion with the industry.

This joint preface was compiled by Oleg Kolosov, with contribution from all the editors of Symposia SS, TT, UU and VV – namely, John Cumings, Dillon D. Fong, Jianyu Huang, Stuart Lindsay, Guangwen Zhou, Jinghua Guo, Hendrik Bluhm, Michael Hävecker, Shu Yamaguchi, Gertjan Koster, Fabio Miletto Granozio, Gyula Eres, Chang-Beom Eom, Nicholas Ingle, Seungbum Hong, Hyunjung Shin and Bryan D. Huey.

John Cumings
Jinghua Guo
Fabio Miletto Granozio
Oleg V. Kolosov

July 2011

ACKNOWLEDGMENTS

We were extremely fortunate to have great invited speakers who presented cutting edge research in this rapidly developing and indispensable area for material science research. We would like to express our gratitude to our Invited Speakers for the high scientific level and novelty of their contribution:

- Fleming Besenbacher, Dawn Bonnell, Robert Carpick, Yasuo Cho, Donna Hurley, Sergei Kalinin, Hyoungsoo Ko, Xiaodong Li, Sergei Magonov, Mervyn Miles, Fabian Mohn, Arvind Narayanaswamy, Christine Ortiz, Sang-il Park, John Pethica, Kumar Wickramasinghe

We also would like to thank Members of the Panel on Next-Generation SPM held at Symposium VV, namely,

- Robert Cook, Kevin Kjoller, John Pethica, Roger Proksch, Sang-il Park, Chanmin Su, Jane Zhu with particular thanks to the panel chair Dawn Bonnell.

The organizers would like to thank the speakers, session chairs, and the symposium assistants for their contribution to make the symposia a success. The organizers are grateful to the MRS editorial staff for their diligent work in organizing the conference and assembling the volume.

And, finally, we gratefully acknowledge the valuable financial support of the Symposia that allowed us to sponsor students and young researchers' participation:

- Asylum Research
- INOSTEK Inc.
- K-TEK Nanotechnology
- Park Systems Corporation
- Bruker Nano Surfaces Business
- SPECS Surface Nano Analysis GmbH
- InSync, Inc.

MATERIALS RESEARCH SOCIETY SYMPOSIUM PROCEEDINGS

Volume 1275— Structural and Chemical Characterization of Metals, Alloys and Compounds, R. Pérez Campos, A. Contreras Cuevas, R.A. Esparza Muñoz, 2011, ISBN 978-1-60511-252-7
Volume 1276— Advanced Structural Materials—2010, H.A. Calderon, A. Salinas Rodriguez, H. Balmori-Ramirez, 2010, ISBN 978-1-60511-253-4
Volume 1277E—Biomaterials—2010, S.E. Rodil, A. Almaguer-Flores, K. Anselme, 2010, ISBN 978-1-60511-254-1
Volume 1278E—Composite, Hybrid Materials and Ecomaterials, R. Bernal, C. Cruz Vazquez, L.E. Rendon Diaz Miron, V.M. Castaño, 2010, ISBN 978-1-60511-255-8
Volume 1279— New Catalytic Materials, J.A. Wang, J. Manuel Domínguez, 2010, ISBN 978-1-60511-256-5
Volume 1280E—Nanomaterials for Biomedical Applications, L. Zhang, T.J. Webster, A. Salinas Rodriguez, 2010, ISBN 978-1-60511-257-2
Volume 1282— Diamond Electronics and Bioelectronics—Fundamentals to Applications IV, P. Bergonzo, J.E. Butler, C.E. Nebel, M. Nesladek, A.T.S. Wee, 2011, ISBN 978-1-60511-259-6
Volume 1283E—Carbon-Based Electronic Devices—Processing, Performance and Reliability, M. Chhowalla, R.R. Keller , M. Meyyappan, W.J. Ready, 2011, ISBN 978-1-60511-260-2
Volume 1284— Fundamentals of Low-Dimensional Carbon Nanomaterials, J.J. Boeckl, L. Dai, W. Lu, M.H. Rümmeli, J. Warner, 2011, ISBN 978-1-60511-261-9
Volume 1285E—Challenges in Roll-to-Roll (R2R) Fabrication for Electronics and Other Functionalities, T. Blaudeck, G. Cho, J.H. Daniel, M.R. Dokmeci, 2011, ISBN 978-1-60511-262-6
Volume 1286E—Molecular and Hybrid Materials for Electronics and Photonics, J. Liu, 2011, ISBN 978-1-60511-263-3
Volume 1287E—Low-Temperature-Processed Thin-Film Transistors, E. Fortunato, 2011, ISBN 978-1-60511-264-0
Volume 1288E—Novel Fabrication Methods for Electronic Devices, P. Andrew, 2011, ISBN 978-1-60511-265-7
Volume 1289E—Controlling Material Properties and Charge-Carrier Interactions with Quantum-Dot Coupling, 2011, ISBN 978-1-60511-266-4
Volume 1290E—Magnetism and Correlated Electronic Structure of Nitrides—Rare-Earth and Transition Metals as Constituents and Dopants, W.R.L. Lambrecht, A. Ney, K. Smith, H.J. Trodahl, 2011, ISBN 978-1-60511-267-1
Volume 1291E—Integrated Nonreciprocal Photonics—Materials, Phenomena and Devices, V. Fratello, M. Levy, B. Stadler, M. Vanwolleghem, 2011, ISBN 978-1-60511-268-8
Volume 1292— Oxide Nanoelectronics, H. Hwang, J. Levy, P. Makysymovych, G. Medeiros-Ribeiro, R. Waser, 2011, ISBN 978-1-60511-269-5
Volume 1293E—Liquid-Crystal Materials—Beyond Displays, N.L. Abbott, D.J. Broer, T. Kato, T.J. White, 2011, ISBN 978-1-60511-270-1
Volume 1294E—Resonant Optical Antennas—Sensing and Shaping Materials, K.B. Crozier, N. Engheta, G. Ju, R. Quidant, R. Zia, 2011, ISBN 978-1-60511-271-8
Volume 1295— Intermetallic-Based Alloys for Structural and Functional Applications, M. Palm, B. Bewlay, S. Kumar, K. Yoshimi, 2011, ISBN 978-1-60511-272-5
Volume 1296E—New Methods in Steel Design—Steel Ab Initio, Y. Adachi, R. Dronskowski, D. Raabe, P.E.A. Turchi, 2011, ISBN 978-1-60511-273-2
Volume 1297— Deformation Mechanisms, Microstructure Evolution and Mechanical Properties of Nanoscale Materials, J.R. Greer, D.S. Gianola, B.G. Clark, T. Zhu, A.H.W. Ngan, 2011, ISBN 978-1-60511-274-9
Volume 1298— Advanced Materials for Applications in Extreme Environments, T.S. Byun, R. Smith, M. Li, 2011, ISBN 978-1-60511-275-6
Volume 1299— Microelectromechanical Systems—Materials and Devices IV, M.P. de Boer, F.W. DelRio, C. Eberl, E.P. Gusev, 2011, ISBN 978-1-60511-276-3
Volume 1300E—Bulk Metallic Glasses and their Applications, K.F. Yao, 2011, ISBN 978-1-60511-277-0

Materials Research Society Symposium Proceedings

Volume 1301— Soft Matter, Biological Materials and Biomedical Materials—Synthesis, Characterization and Applications, A.J. Nolte, K. Shiba, R. Narayan, D. Nolte, 2011, ISBN 978-1-60511-278-7
Volume 1302E—Nanowires—Growth and Device Assembly for Novel Applications, 2011, ISBN 978-1-60511-279-4
Volume 1303— Nanomaterials Integration for Electronics, Energy and Sensing, D. E. Perea, Y. Jung, J. B. Hannon, M. A. Reed, S. T. Picraux, 2011, ISBN 978-1-60511-280-0
Volume 1304E—Hierarchical Materials and Composites—Combining Length Scales from Nano to Macro, J.H. Moon, G.M. Odegard, M.S.P. Shaffer, B.L. Wardle, 2011, ISBN 978-1-60511-281-7
Volume 1305E—Group IV Semiconductor Nanostructures and Applications, L. Dal Negro, 2011, ISBN 978-1-60511-282-4
Volume 1306E—Aerogels and Aerogel-Inspired Materials, S. Brock, G. Gould, A. Roig, D. Rolison, 2011, ISBN 978-1-60511-283-1
Volume 1307E—Boron and Boron Compounds—From Fundamentals to Applications, M. Dudley, J.H. Edgar, M. Kuball, 2011, ISBN 978-1-60511-284-8
Volume 1308E—Artificially Induced Crystalline Alignment in Thin Films and Nanostructures, A.T. Findikoglu, R. Huehne, T. Shimada, J.Z. Wu, 2011, ISBN 978-1-60511-285-5
Volume 1309— Solid-State Chemistry of Inorganic Materials VIII, K-S. Choi, S.J. Clarke, P.S. Halasyamani, D.G. Mandrus, 2011, ISBN 978-1-60511-286-2
Volume 1310E—Magneto Calorics and Magnetic Cooling, A. Fujita, K. Gschneidner Jr., O. Gutfleisch, K.G. Sandeman, A. Yan, 2011, ISBN 978-1-60511-287-9
Volume 1311— Next-Generation Fuel Cells—New Materials and Concepts, T. He, K. Swider-Lyons, B. Park, P.A. Kohl, 2011, ISBN 978-1-60511-288-6
Volume 1312— Polymer-Based Materials and Composites—Synthesis, Assembly, Properties and Applications, V. Bharti, M. Chipara, D. Venkataraman, 2011, ISBN 978-1-60511-289-3
Volume 1313— Materials for Advanced Lithium Batteries, G.-A. Nazri, J-M Tarascon, D. Guyomard, A. Yamada, 2011, ISBN 978-1-60511-290-9
Volume 1314E—Thermoelectric Materials for Solid-State Power Generation and Refrigeration, Y. Grin, G.S. Nolas, J. Sharp, T.M. Tritt, 2011, ISBN 978-1-60511-291-6
Volume 1315— Transparent Conducting Oxides and Applications, J.J. Berry, E. Fortunato, J. Medvedeva, Y. Shigesato, 2011, ISBN 978-1-60511-292-3
Volume 1316E—Nanofunctional Materials, Nanostructures and Nanodevices for Biomedical Applications II, R. Rao, 2011, ISBN 978-1-60511-293-0
Volume 1317E—Interdisciplinary Approaches to Safe Nanotechnologies, C. Chaneac, S. Harper, G.V. Lowry, R.I. MacCuspie, 2011, ISBN 978-1-60511-294-7
Volume 1318— Advances in Spectroscopy and Imaging of Surfaces and Nanostructures, J. Cumings, J. Guo, F.M. Granozio, O.V. Kolosov, 2011, ISBN 978-1-60511-295-4
Volume 1319— Materials Issues in Art and Archaeology IX, P.B. Vandiver, C.L. Reedy, J.L. Ruvalcaba Sil, W. Li, 2011, ISBN 978-1-60511-296-1
Volume 1320— Materials Education Development and Outreach—From K–Grad, D. Bahr, K. Jones, M. Glass, E. Allen, 2011, ISBN 978-1-60511-297-8

Prior Materials Research Society Symposium Proceedings available by contacting Materials Research Society

Advanced Imaging and Scattering Techniques for In Situ Studies

Mater. Res. Soc. Symp. Proc. Vol. 1318 © 2011 Materials Research Society
DOI: 10.1557/opl.2011.917

In situ Imaging at the NIST Neutron Imaging Facility

David L. Jacobson, Daniel S. Hussey and Eli Baltic
Physical Measurement Laboratory, National Institute of Standards and Technology, Gaithersburg, MD 20899-8461, U.S.A.

ABSTRACT

Neutron imaging as a method to perform *in situ* studies of hydrogen fuel cells, hydrogen storage devices, heat pipes, and batteries has made tremendous progress in recent years. Neutrons are useful to study light elements mixed with heavy Z elements where penetration by other forms of radiation is either impossible or incapable of contrasting the light elements. Useful spatial resolution available at neutron imaging facilities is now approaching 10 micrometers. Complimentary time resolution of 30 fps or greater is also possible with a spatial resolution approaching 300 micrometers. Here we will provide an overview of the technique of neutron imaging and experimental studies with neutrons at the National Institute of Standards and Technology. Examples of *in situ* studies of fuel cells, hydrogen storage devices, heat pipes and batteries will be discussed.

INTRODUCTION

The method of neutron radiography has seen a strong resurgence in recent years due mainly to the advent of digital imaging detectors in the 1990's. This resurgence is also due in part to the unique ability of the neutron to penetrate metal components and remain sensitive to light elements like hydrogen and lithium that are found in important alternative energy storage and conversions devices. Examples include hydrogen powered polymer electrolyte membrane fuel cells (PEMFC), hydrogen storage beds and alkaline and lithium- ion batteries. As the neutron digital imaging field has progressed, the spatial resolution has been dramatically improved from 250 μm to the current best case measured resolution of 13 μm.

Neutron radiography is best implemented at a research reactor facility which can provide enough neutron intensity to allow for both dynamic imaging (greater than or equal to 30 fps) or high resolution imaging (near 10 μm). The NIST Neutron Imaging Facility (NNIF) offers users the opportunity to utilize neutrons for imaging [1,2].

A raw neutron image is called a radiograph and is a 2-dimensional image of the beam intensity that passes through an experimental device. The resulting image is simply a shadow or transmission image of the device. Regions of greater varying attenuation are described well by the Lambert-Beer relationship given by:

$$I = I_0 e^{-N\sigma t} \tag{1}$$

where N is the atom density, σ is the neutron cross section of the atom, and t is the thickness of material that the neutron beam passes through. In some cases it is possible to quantify how much material the neutron passes through using Eq. (1). Shown in figure 1 is σ for selected elements, which shows that the interaction between neutrons and material does not depend on the atomic number, Z. This is due to the fact that the neutron interacts with the nucleus through the strong nuclear force and not with the Coulombic forces in the atom. From figure 1 it is apparent that the cross sections for numerous high Z materials are significantly smaller than the cross section for H and Li. This means that for devices such as fuel cells, hydrogen storage beds, heat pipes, and batteries that are housed in Al or thin stainless steel housings, little or no modification to the device has to be made in order to image the device and it's contents.

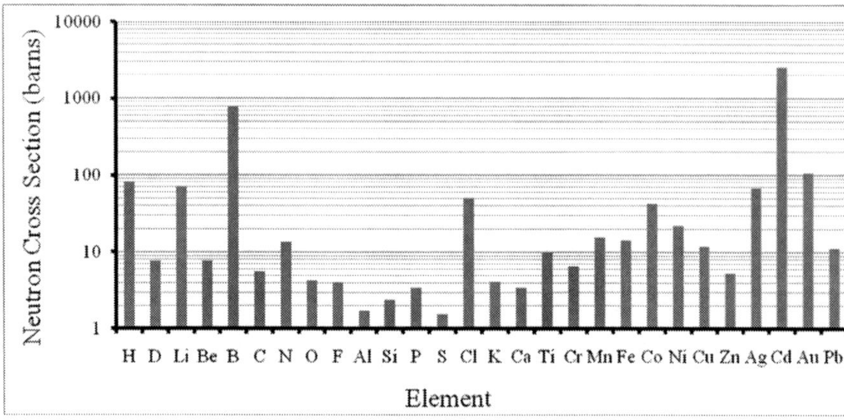

Figure 1. Neutron cross sections of selected elements.

The next most important aspect of imaging with neutrons is the spatial resolution of the images that are formed. The blurring in the image is the combined blurring due to both the beam and the finite resolution of the detector. The full width at half maximum (FWHM) of the blurring from the beam is known as the geometric un-sharpness, δ_g, and can be calculated from the L/D ratio (depicted in figure 2), along with the sample-to-detector separation, z, using the following relation,

$$\delta_g = \frac{zD}{(L-z)} \approx zD/L \tag{2}$$

The blurring can be modeled as a Gaussian point spread function given by

$$PSF(x-x') = \frac{1}{\sqrt{2\pi\sigma^2}} \exp\left(-\frac{(x-x')^2}{2\sigma^2}\right) \tag{3}$$

where the FWHM is related to the Gaussian standard deviation by

$$\delta = \frac{\sigma\pi}{\sqrt{2\ln(10)}} \tag{4}$$

When modeling the blurring due to the detector and the L/D ratio as a Gaussian blur the two terms can be combined in quadrature to calculate the overall FWHM given by

$$\delta = \sqrt{\delta_d^2 + \delta_g^2} \qquad (5)$$

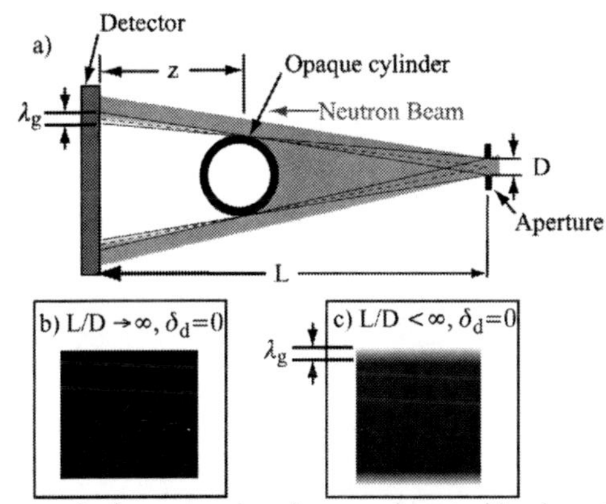

Figure 2. Standard layout a) of a neutron imaging experiment with relevant dimensions noted. Ideal resolution shown in b) for no detector blur. Geometric blur shown in c) resulting from a finite aperture size D.

EXPERIMENT

The NNIF is located at the NIST Center for Neutron Research. Here the center provides a high flux 20 MW reactor-based neutron source of thermal neutrons for experiments. The facility, shown in figure 3, is a national user facility that provides at least 25 % of the available beam time to users through a peer reviewed call for proposals. Nearly any L/D ratio may be achieved at this facility, but as a figure of merit the limiting L/D ratio for L = 6 m is 300. With this L/D ratio the neutron fluence rate is 3×10^7 after being filtered for gammas and fast neutrons with a liquid nitrogen cooled Bi filter that is 10 cm in length. The resulting beam size at the sample position is 26 cm in diameter. The beam intensity roughly scales as

$$I_2 = I_1 \left(\frac{L_1 D_2}{D_1 L_2} \right)^2, \qquad (6)$$

which connects the intensity at L_1/D_1 ratio with a second intensity at another L_2/D_2 ratio. Samples can be brought into the facility and mounted on sample manipulation stages located at position 8) in the schematic shown in figure 3. By rotating the object by 180° it is possible to perform 3-dimensional tomographic reconstructions of the object being studied.

The imaging facility has several types of digital detector systems that allow 2-dimensional images of the neutron beam to be captured. The first detector system is a scintillator viewed by a charge-coupled-device (CCD) camera. The best spatial resolution achieved to date with the current system is 18 μm for a pixel size of 8.5 μm. Each pixel is

8.5 μm square and there are 1002 by 1004 pixels, which means that the field of view at this spatial resolution is about 0.85 cm square. To achieve larger fields of view the pixel count of the CCD must be increased or the minification ratio of the lens or must be adjusted, but this will also increase the value of the spatial resolution.

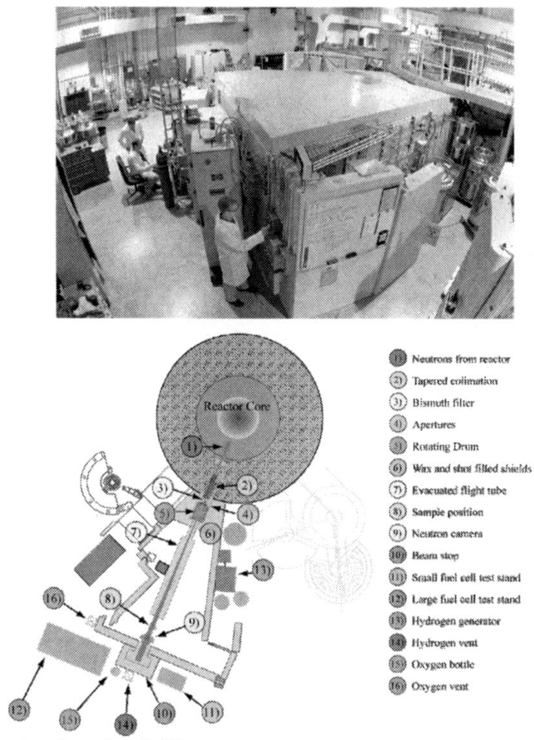

Figure 3. NIST neutron imaging facility.

The second detector system is an amorphous silicon detector, wherein a scintillator is coupled directly to an amorphous silicon panel without a lens. This means that the detector has a fixed spatial resolution of 254 μm with 127 μm square pixels and fixed field of view; but by directly coupling the scintillator to the sensor, this system is optimized for capturing light and can therefore capture images at a rate up to 30 fps.

The third detector system is a micro-channel plate (MCP) neutron detector [3,4]. Instead of a scintillator, the charged particles from the nuclear reaction in the MCP produce an electron avalanche that is accelerated out of the MCP via an applied electric field onto a cross strip anode which performs determines the position of the neutron capture via a centroiding algorithm. The pixel size of this detector is 5 μm square pixels and the measured spatial resolution is 13 μm with a maximum viewable area of 4.0 cm in diameter.

DISCUSSION

The sensitivity to hydrogen and lithium lends neutron imaging to the study of devices that are composed of metal with internal parts that are hydrogenous or lithiated. Experiments incorporating devices of technical significance such as hydrogen fuel cells, hydrogen storage devices, heat pipes and batteries have been of interest to researchers who have made use of the facility.

Fuel Cells

Polymer Electrolyte Membrane (PEM) Fuel Cells electrochemically combine hydrogen and oxygen thereby producing electricity with only non-polluting byproducts of heat and water. Neutron radiography can be used to accurately map both the distribution and quantity of water in the fuel cell (figure 4) [5-16]. From these images researchers have been able to both model and

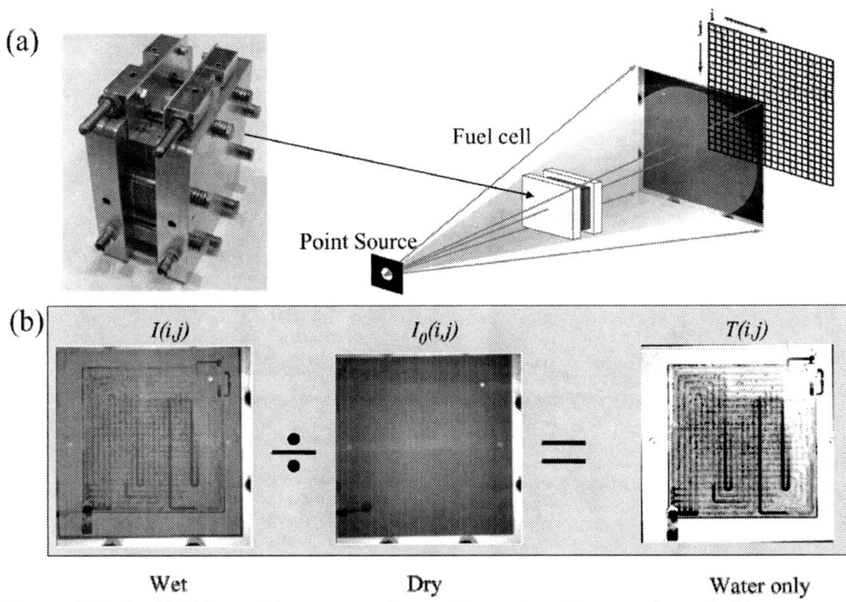

Figure 4. Fuel cell (a) imaged by a neutron beam. Processing of images shown in (b) to extract quantitative image of water distribution.

experimentally study the effects of heat and mass transport on the water distribution in the fuel cell. During operation, humidified gases are fed to the fuel cell that operates at a nominal temperature of 80 °C, which is why liquid water exists in the cell. In addition to the water produced during the chemical reaction of hydrogen with oxygen, water can also come from condensation from the input humidified gases as the carrier gas/fuel is consumed. When run in cold climates where ambient temperatures drop below the freezing point of water, any residual water can be a significant source of degradation of fuel cell performance and premature failure. With the recent improvement of spatial resolution to 10 μm more fundamental studies of water transport effects from the anode to cathode - such as thermally driven transport - are being performed.

Batteries

Neutron imaging plays a similar role in advanced battery development by providing *in situ* measurements of hydrogen or lithium distributions at different states of charge of the battery[17-20]. In alkaline primary cells, the neutron attenuation is dominated by the aqueous electrolyte. Tomograms (three-dimensional images) of two AA alkaline cells were acquired before and after discharge to a cell potential of 0 V. Shown in figure 5 are tomographic slices through two different AA batteries at the end of life achieved by two different current draw conditions, 50 mA and 1 A. In the initial state, the distribution of electrolyte is homogenous in the anode and the cathode. Also, the separator, which is composed of a porous hydrogenous material such as rayon where the pores are filled with electrolyte solution, is clearly visible and is highly attenuating to neutrons. The distribution of the electrolyte after the cell potential has fallen to 0 V clearly depends on the discharge rate. By optimizing particle size to

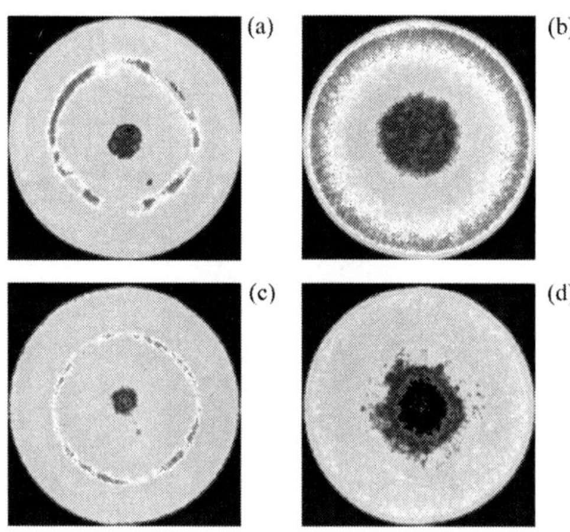

Figure 5. Comparison of the changes in an alkaline AA cell after operation at two different constant currents, where black is low neutron attenuation, red is high neutron attenuation. (a) Fresh cell before operation at 50 mA, (b) discharged cell after operation at 50 mA for 52.5 h, (c) fresh cell before operation at a constant 1 A, (d) discharged cell after operation at 1 A for about 70 min. The electrolyte has an initially uniform distribution (green) in the anode (interior) and in the cathode (exterior), and the separator (red/yellow) is clearly visible in (a) and (c).

reduce the removal of electrolyte from the anode, it is possible to increase the cell capacity. Shown in figure 6 is the evolution of a lithium primary cell showing the collapse of the jelly roll into the inner void space as the battery is discharged. In more recent experiments, two-dimensional neutron imaging has been used to study a wound prismatic lithium-ion battery. By using high resolution neutron imaging, the thermal expansion of the battery on charge and discharge can be directly measured.

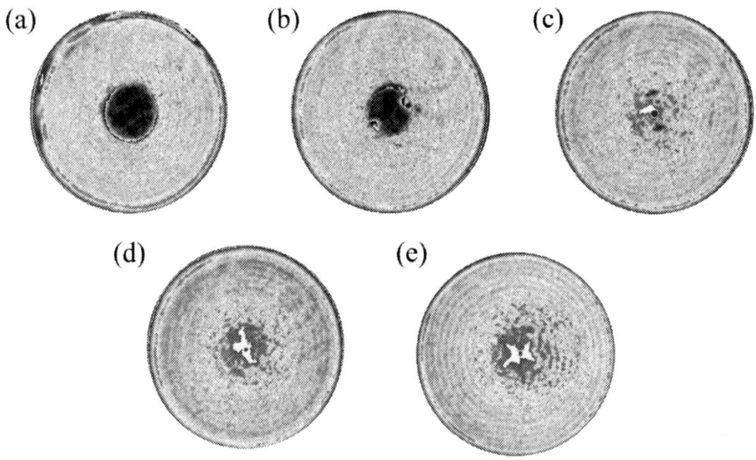

Figure 6. Summary of lithium primary cell tomography. These data have been corrected for beam hardening. (a) Fresh cell, (b) after 1 A h discharge at 1 A, (c) after 2 A h discharge at 1 A, (d) after 3 A h discharge at 1 A (e) after 3 A h discharge at 0.1 A.

Hydrogen Storage

Future hydrogen fuel cell vehicles will require hydrogen storage vessels that efficiently store and quickly release the hydrogen fuel. The hydrogen uptake in most storage materials strongly depends on the thermal environment, thus optimizing the fuel tank will require understanding the coupled heat and mass transport system. While many techniques exist to measure the volume average of the hydrogen uptake, to date no method exists that can measure the full three-dimensional hydrogen distribution within the storage vessel. In order to address this shortcoming, a prototype hydrogen storage bed has been studied using neutron tomography and radiography [21].

The prototype metal hydride bed consisted of a ≈90 % porous aluminum foam filled with $LaNi_{4.78}Sn_{0.22}$ powder. The foam provided structural integrity and enhanced the thermal conductivity of the system. A heater rod with four turns ran the length of the bed and was in contact with the aluminum foam to heat the metal hydride during hydrogen desorption. Hydrogen gas was introduced into and removed from the bed via a hollow stainless steel filter tube in the center. One aspect of this study, shown in figure 7, involved acquiring two tomograms of the hydrogen storage bed: the first was after the bed was heated to 100 °C and evacuated to a pressure below 10^{-6} mbar; the second was after the bed was charged with about 12 standard liters of hydrogen.

Non-uniform hydrogen distributions were observed during this study and were attributed to the non-uniform distribution of temperature in the bed. Hydrogen absorption is an exothermic process, and the hydrogen uptake decreases with increasing temperature. The study observed that hydrogen is seen to preferentially absorb along the circumference and accumulate towards the end of the storage bed. This was attributed to the absorption of hydrogen by $LaNi_{4.78}Sn_{0.22}$ being exothermic coupled with reduced hydrogen uptake at higher temperatures. From neutron imaging an accurate picture of the hydrogen distribution in a storage bed can be obtained and compared with models of heat and mass transfer in order to optimize the storage and transport of hydrogen in the bed.

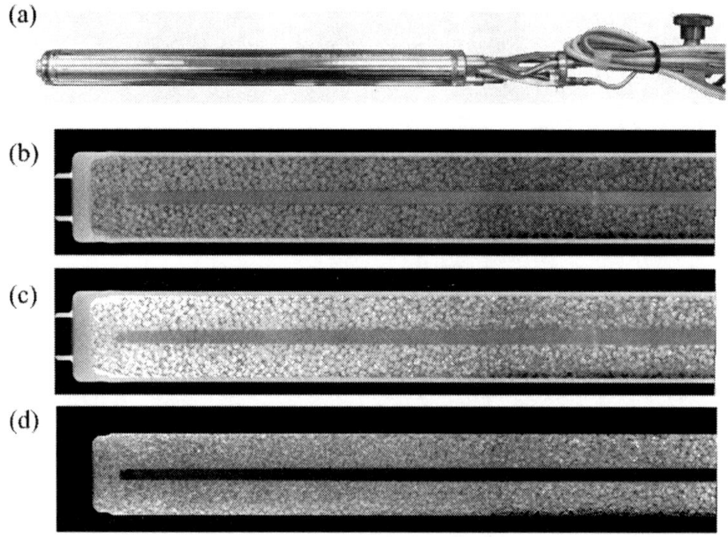

Figure 7. Picture of the (a) Plank hydrogen storage bed. Sagittal slices through the tomographic reconstructions of the bed for (b) empty, (c) the bed filled with 94.5 L of D2 and (d) the hydrogen only after subtracting (b) from (c).

Heat Pipes

One of many applications for heat pipes have been in modern computer systems to move heat away from components such as CPUs and GPUs to heat sinks where thermal energy may be dissipated into the environment. Heat pipes have no mechanical components. Instead they are closed systems of liquid and vapor phases that rely on the process of vaporization and condensation to drive the system. Liquid absorbs heat in the evaporator and vaporizes. The vapor expands and pushes liquid from the connected condenser back into a second evaporator.

The newly created vapor then expands into the condenser where it cools rejecting heat back into the local environment and condensing back into the liquid state. Meanwhile the liquid driven into the second evaporator vaporizes. This second vaporization then drives the liquid (from condensed vapor) back to the first evaporator where the cycle starts all over again. This cycle or oscillation continues back and forth as long as heat is being generated (in this case the heat is created by a CPU). This system is known as an Oscillating Heat Pipe (OHP). Previous experiments performed to visualize the oscillation have had to modify the metal pipe system to use glass or plastic to provide a visible window. Since these systems are all usually made of copper, the use of glass or plastic changes the thermal transfer properties of the system in turn affecting the performance of the system. Neutron imaging allows researchers the opportunity to study the liquid flow in these systems *in situ* using the actual hardware. Due to the high intensity of neutron beam available at the NNIF it is possible to quickly capture dynamic images at 1/30th of a second. To do this the facility utilizes an advanced amorphous silicon imaging system that is capable of capturing neutron images at this extremely high rate. Wilson et al [22] utilized this capability to visualize the effects of the filling fraction of the heat pipe on the fluid distribution and coupled this with thermal evaluations to study the consistency of the heat transport and oscillation in the flow of nanofluids through heat pipes.

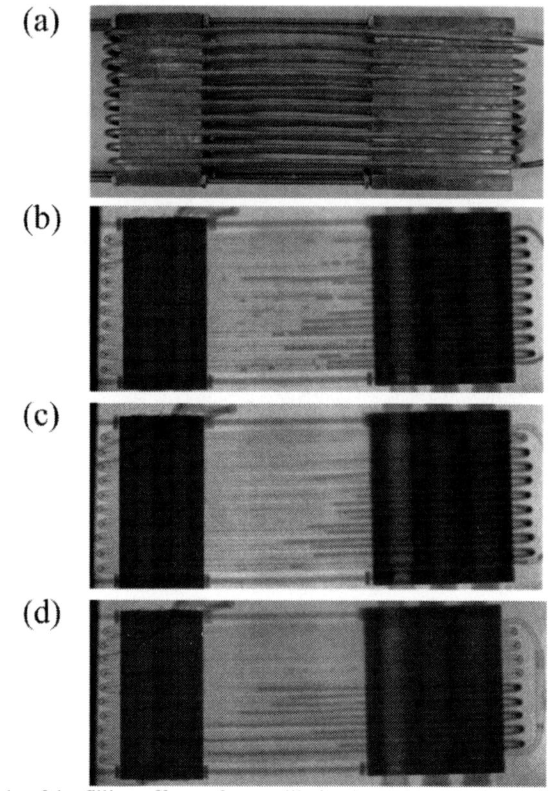

Figure 8. A study of the filling effects of an oscillating heat pipe, (a) photo of the 10-turn heat pipe, typical fluid distribution at 50 W applied heat for filling fraction (b) 53 %, (c) 35 % and (d) 30 %.

CONCLUSIONS

Due to the unique ability to penetrate metals and other high Z materials and maintain high contrast to H and Li, neutron radiography and tomography are tools that allow one to study devices that are not easily studied with other methods like x-ray radiography. Advances in imaging technology have and will continue to provide new avenues of research for neutron imaging for some time to come. Current detection limits of 10 μm may be pushed even lower to near 1 μm in the future. Also, faster conducting metal oxide semiconductor (CMOS) based sensors may allow for frame rates up to 100 fps, which will further extend the dynamic image capture capabilities to study heat pipe phenomenon. In the last decade neutron radiography has made significant contributions to the understanding of hydrogen fuel cells. The potential for neutron radiography to make impacts in hydrogen storage device applications, heat pipe and battery research has also been demonstrated.

ACKNOWLEDGMENTS

This work was supported by the U.S. Department of Commerce, the NIST Physical Measurement Laboratory, the Director's office of NIST, the NIST Center for Neutron Research, and the Department of Energy interagency agreement Nos. DE-AI-01-01EE50660 and DE-AI-01-05EE11104.

REFERENCES

1. D. L. Jacobson, D. S. Hussey, E. Baltic, J. LaRock, M. Arif, J. Gagliardo, J. P. Owejan, T. A. Trabold in *Neutron radiography, proceedings of the 8th world conference on neutron radiography,* edited by M. Arif M, R.G. Downing, (DES Tech Publications, Lancaster, PA, 2008) pp. 50-57.
2. D. S. Hussey, D. L. Jacobson, M. Arif, P. R. Huffman, R. E. Williams, J. C. Cook, *Nucl. Instr. Meth.* **A542**, 9-15 (2005).
3. O. H. W. Siegmund, J. V. Vallerga, A. Martin, B. Feller, M. Arif, D. S. Hussey, D. L. Jacobson, *Nucl. Instr. Meth.* **A579**, 188-191 (2007).
4. A. S. Tremsin, J. V. Vallerga, J. B. McPhate, O. H. W. Siegmund, W. B. Feller, L. Crow, R. G. Cooper, *Nucl. Instr. Meth.* **A592**, 374-384 (2008).
5. R. J. Bellows, M. Y. Lin, M. Arif, A. K. Thompson, D. Jacobson, *J. Electrochem. Soc.* **146**, 1099-1103 (1999).
6. R. Satija, D. L. Jacobson, M. Arif, S. A. Werner, *J. Power Sources* **129**, 238-245 (2004).
7. I. A. Schneider, D. Kramer, A. Wokaun, G. G. Scherer, *Electrochem. Commun.* **7**, 1393-1397 (2005).
8. T. A. Trabold, J. P. Owejan, J. J. Gagliardo, D. L. Jacobson, D. S. Hussey, M. Arif in *Handbook of fuel cells: advances in electrocatalysis, materials, diagnostics and durability* edited by W. Vielstich, H.A. Gasteiger, H. Yokokawa, (Handbook of fuel cells **5** & **6**, John Wiley & Sons Ltd., Chichester, UK, 2009) pp. 658-672.
9. D. S. Hussey, D. L. Jacobson, M. Arif, J. P. Owejan, J. J. Gagliardo, T. A. Trabold, *J. Power Sources* **172**, 225-228 (2007).
10. P. Boillat, D. Kramer, B. C. Seyfang, G. Frei, E. Lehmann, G. G. Scherer, A. Wokaun, Y. Ichikawa, Y. Tasaki, K. Shinohara, *Electrochem. Commun.* **10**, 546-550 (2008).
11. M. A. Hickner, N. P. Siegel, K. S. Chen, D. S. Hussey, D. L. Jacobson, M. Arif, *J. Electrochem. Soc.* **155(3)**, B294-B302 (2008).
12. A. Turhan, K. Heller, J. S. Brenizer, M. M. Mench, *J. Power Sources* **180**, 773-783 (2008).
13. J. P. Owejan, T. A. Trabold, D. L. Jacobson, M. Arif, S. G. Kandlikar, *Int. J. Hydrogen Energy* **32**, 4489-4502 (2007).
14. M. A. Hickner, N. P. Siegel, K. S. Chen, D. S. Hussey, D. L. Jacobson, M. Arif, *J. Electrochem. Soc.* **155**, B427-B434 (2008).
15. S. Kim, M. M. Mench, *J. Electrochem. Soc.* **156**, B353 (2009).
16. D. Spernjak, S. G. Advani, and A. K. Prasad, *J. Electrochem. Soc.* **156**, B109-B117 (2009).
17. M. Kamata, T. Esaka, K. Takami, S. Takai, S. Fujine, K. Yoneda, K. Kanda, Solid State Ionics 91, 303-306 (1996).

18. M. Lanz, E. Lehmann, R. Imhof, I. Exnar, P. Novak, *J. Power Sources* **101**, 177-181 (2001).
19. N. Kardjilov, A. Hilger, I. Manke, M. Strobla, W. Treimer, J. Banhart, *Nucl. Instr. Meth.* **A542**, 16–21, (2005).
20. G. V. Riley, D. S. Hussey, D. L. Jacobson, ECS Trans. 25, 75 (2010).
21. D. L. Jacobson, D. S. Hussey, E. Baltic, T. J. Udovic, J. J. Rush, R. C. Bowman, Jr., *Int. J. Hydrogen Energy* **35**, 12837–12845 (2010).
22. C. Wilson, B. Borgmeyer, R. A. Winholtz, H. B. Ma, D. L. Jacobson, D. S. Hussey, M. Arif, *J. Thermophys. Heat Transf.* **22**, 366-372 (2008).

Mater. Res. Soc. Symp. Proc. Vol. 1318 © 2011 Materials Research Society
DOI: 10.1557/opl.2011.918

Low Energy Ne Scattering Spectroscopy for Insulators, and Materials in the Electric/Magnetic Fields

Kenji Umezawa[1], Shigemitsu Nakanishi[2], Hideki Hayashi[2], Hideaki Higashitsutsumi[2], Hiroki Nagasawa[2], Eisuke Narihiro[3], Keiko Ogai[3]
[1] Dept. of Physics, Osaka Prefecture University, 1-1 Gakuen-Cho, Sakai 599-8531, Japan
[2] Pascal Co. Ltd., 1-1-4 Shouwa-Cho, Abeno, Osaka 545-0011, Japan
[3] Apco Co. Ltd., 522-10 Kitano-Cho, Hachiouji, Tokyo 192-0906, Japan

ABSTRACT

This study describes a low-energy atom scattering system that was combined with a time-of-flight spectrometer for insulator surface structural analysis. We show one example. MgO(001) crystal was used to study the surface analysis technique and is illustrated here. Insulator surface structure is difficult to study because of the charging effects during electron or ion-beam bombardment. Nevertheless, structural analysis of insulator surfaces is very important in fundamental research as well as in technology fields.

INTRODUCTION

Surface science is present in many important technologies and processes and is the basis for everyday applications, such as electronic devices. Fundamental concepts for surface science are well-established and their applications are straightforward for metals and semiconductors, but not for insulators and materials in the electric/magnetic fields. Bombardment of insulator surfaces by charged ions can be induced a charge on their surfaces. One can see the charging/discharging dynamics of the insulating material during this ion-beam bombardment. Sometimes, an electron shower using a tungsten filament placed nearby a sample is used to reduce the sample charging. However, electron-shower failure can cause sample damage.

Therefore, we developed a low-energy atom scattering spectroscopy system for the analysis of these insulator surfaces. Low-energy atom beams were produced using ion beams. Atom beams were converted from ion beams by charge exchange while the ion beams passed through a small gas chamber (gas is the same element as ion beams.).

In this study, low-energy ion scattering spectroscopy was modified for the analysis of insulator surfaces. Low-energy (less than 5 keV) ion scattering technique is quite useful for the analysis of semiconductor as well as metal surfaces [1-5]. Low-energy atom particle beams were projected onto the sample surfaces to avoid charging effects. Measurements were performed by the time-of-flight technique. (Low-keV ions have a short de Broglie wavelength (~ 10^{-14} m), which constrains them to classical behavior, as well as large collision cross-sections. In the energy region between 1 and 5 keV of He+ /

Ne^+ beams, the Thomas-Fermi-Moliere (TFM) potential or Ziegler-Biersack-Littmark (ZBL) was used for the analysis [6].)

EXPERIMENTAL

The low-energy atom scattering spectrometer systems for surface structural analysis of insulators were constructed from both custom-made and commercially available components. Almost all of the electric devices for the measurements were designed and hand-built by us. Figure 1 shows the equipment of the low-energy atom scattering spectroscopy system, which comprises an ultra high vacuum (UHV) chamber with the primary atom beam line: (1) ion beam source; (2) MCP detector; (3) precision sample manipulator with stepping motors (custom-made); (4) pre-amplifier; (5) amplifier; (6) 4ch multiple-stop time-to-digital converter; (7) pulse generator; 8) low-energy electron diffraction (LEED) optics and Auger Electron Spectroscopy (AES); and (9) software to collect data and control the stepping motor.

A 2 keV-He^+ ion beam is typically produced by an ion source with an electron impact type and research-grade He gas. An ion beam is collimated to a diameter of 1.5 mm by an aperture. The position of the collimated ion beam is adjusted by an electrostatic shifter to pass the right middle of a deflector. The chopping of the ion beam is carried out when the ion beam passes the deflector to produce a pulse beam. The voltage applied to the chopping plates is 50 V with a pulse width less than 10 nsec. The chopping frequency is typically 100 kHz. A pulse beam is neutralized in the small charge exchange cell where He gas was introduced through a variable leak valve up to a pressure of 1×10^{-2} Pa. The conversion efficiency from He^+ to He^0 beam was about 34 %. A neutralized beam passed through an MCP (micro-channel plate) that has a center hole with a diameter of 5 mm. An MCP detector (assembly outer diameter = 34 mm) is coaxially mounted along the primary drift tube. The residual $^4He^+$ ions were removed by the deflector at the entrance of UHV

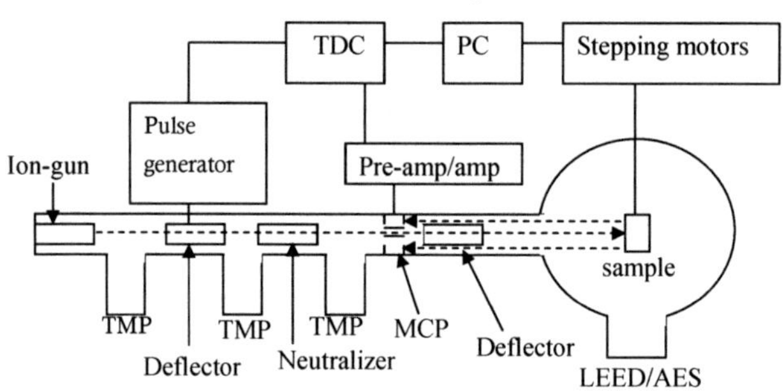

Figure 1 Experimental equipment of low-energy atom scattering spectroscopy.

chamber. Thus, $^{4}He^{0}$ particles impinge on a sample and the scattered particles are detected by an MCP that is located at 180°. The time-of-flight measurements are carried out using a trigger circuit and a time-to-digital converter (TDC) that has four independent channels. Each channel independently measures the time between a trigger and stop signals at a time resolution of 10 nsec. Data acquisition mode (polar scan and azimuth scan) has been automated, and collected in 2.0° steps. Originally, the ion-gun, pulse generator, pre-amp/amp, TDC, and stepping motor systems were homemade equipment.

DISCUSSION

Figure 2 shows a typical example of a series of low-energy ion and atom scattering spectra as a function of flight time. The sample was MgO(001) crystal. Time-of-flight spectra labeled by "a" and "b" were obtained from the primary 2 keV-$^{4}He^{+}$ incident beams, and primary 2 keV-$^{4}He^{0}$ incident beams, respectively. In the beginning of the measurements, the spectrum could be obtained from the primary 2 keV-$^{4}He^{+}$ incident beams, as shown in this figure.

However, it became hard to obtain the spectrum because of charge effects on sample surfaces due to incident ion beams. On the other hand, atom scattering spectrum labeled by "b" could always be obtained as a function of measuring time because of the electron neutrality of the primary incident beams. The other spectra labeled by "c" were obtained from the primary 2 keV-$^{4}He^{+}$ incident beams in an electric field (50 V/cm). The voltage applied to the deflector plate was 50 V. The deflector was located in a beam line.

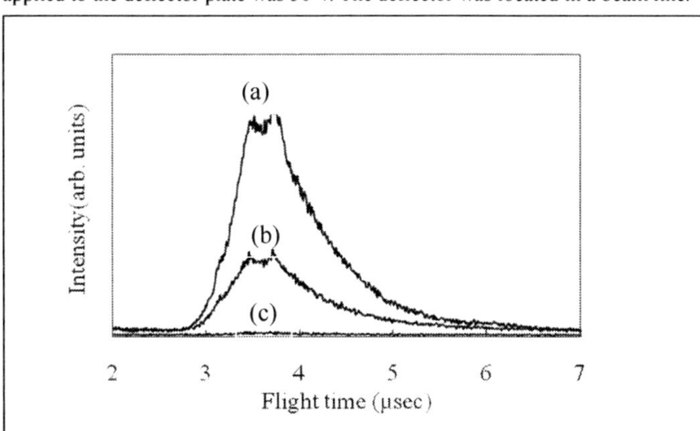

Figure 2 Typical specta of backscattered particles as a function of the flight time. (a) shows the spectra obtained from 2 keV-$^{4}He^{+}$ incident beams. (b) shows the spectra obtained from 2 keV-$^{4}He^{0}$ incident beams. (c) shows the background spectra.

The primary 2 keV-$^4He^+$ beams were basically bent due to electric field and did not hit the sample surfaces. The spectrum labeled by "c" shows the background intensity. One can see that the intensity of the background spectrum is much smaller than the atom scattering spectrum labeled by "b".

Figure 3 shows the variation of the intensity of backscattered particles from MgO(100) sample surfaces as a function of azimuth angles along the polar angle of 30°of the 2 keV-$^4He^0$ particles' incident direction. The intensity drops observed at 0°([100]), ±45°([110], [1-10]), and ±90°([010], [0-10]) show the shadowing effects for the polar angle of 30°, respectively. In addition, the intensity peaks around ±23° and ±68° are due to the focusing effects. These profiles clearly show a periodicity of 90°for azimuthal angle (fourfold symmetry) with respect to the MgO(001) sample surfaces.

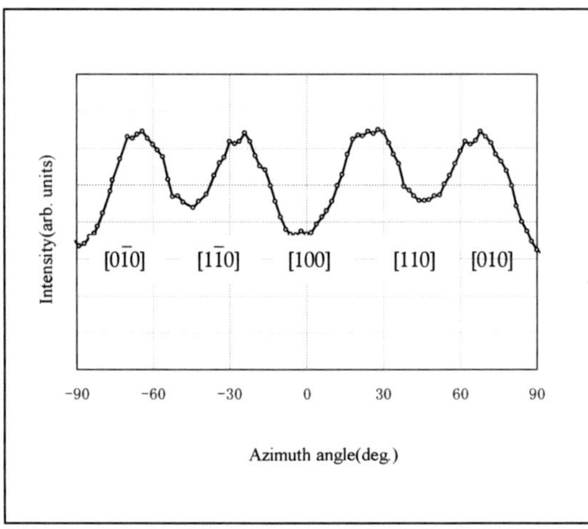

Figure 3 Backscattered particles intensity as a function of the azimuth angles. The primary incident beam was 2 keV-$^4He^0$.

Figure 4 shows the intensity profiles as a function of the polar angle obtained from MgO(001) sample surfaces. Observed peaks shown in this figure can be divided into two groups: one group labeled a, b, c and another labeled a, -b, -c.

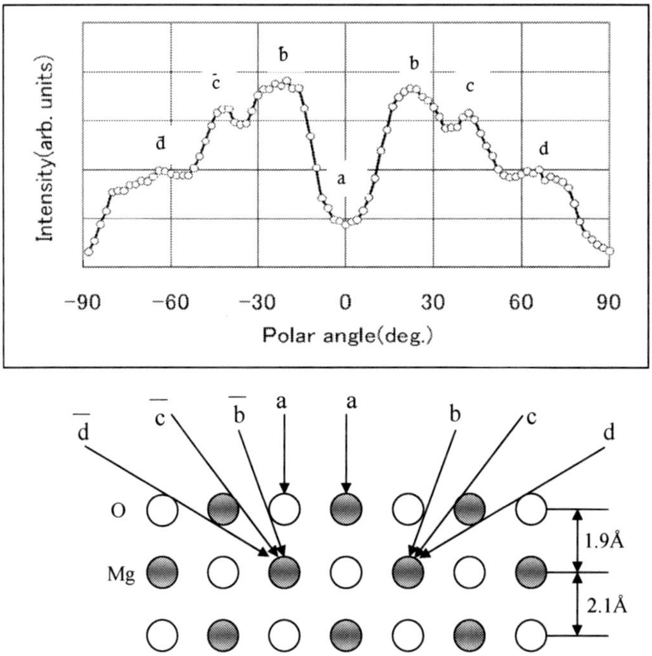

Figure 4 A series of polar scans for 2 keV-$^4He^0$ backscattered from MgO(001) sample surfaces along [100] azimuth. Side view of the MgO(001) planes. The alphabetic symbols are positions where focusing effects are expected.

Symbols shown by "a" indicate signals coming from the first layer. Symbols shown by "b" and "-b," "c" and "-c," and, "d" and "-d," indicate signals coming from the second layer. However, peak positions depend on the incident angle. Although we have not done computer simulations yet, peak positions based on shadowing effects indicate that the spacing between the first layer and the second layer is 1.9 Å. This is smaller than spacing in a bulk (2.1 Å). All of the scattering intensity depends on the incident angles. Due to the influence of the neighboring atoms, peak positions change in the polar angle at which the edge of the shadowcone intercepts adjacent atoms for the particular plane for surface structure. Thus, at particular angles, spacing can be obtained. The Mg image of clean MgO(100) surfaces is shown in Figure 5. It is obtained from rotating the sample. Mg signals were picked up. This figure also shows the crystal structure of (100) surfaces.

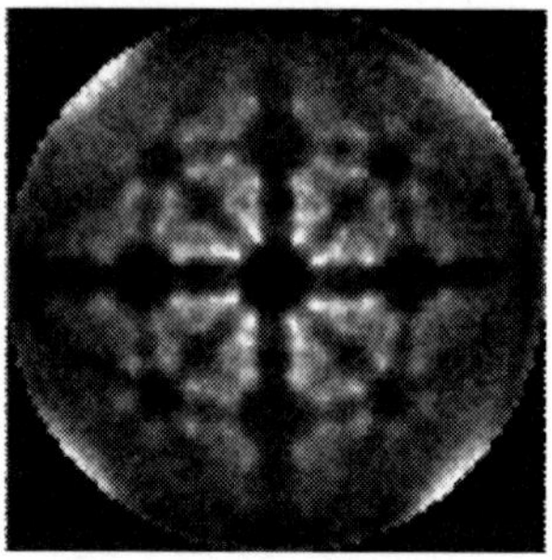

Figure 5 the Mg image of clean MgO(100) surfaces

CONCLUSIONS

This study described a low-energy atom scattering system combined with a time-of-flight spectrometer for insulator surface structural analysis. We show one of examples to study of MgO(001) surface analysis. Insulator surface structure is difficult to study because of charging effects during electron or ion beam bombardment. Structural analyses of insulator surfaces are very important in fundamental research as well as technology fields.

ACKNOWLEDGEMENTS

This research was supported by (1) MEXT, Grants-in-Aid for scientific research on innovative areas between the fiscal years from 2008 to 2010, (2) JSPS, Grants-in-Aid for scientific research (C) between the fiscal years from 2008 to 2010, (3) JST-modeling of innovative research results in 2009, and (4) JST-project to develop "innovative seeds (the equipment : TOFLAS-3000)" between the fiscal years from 2008 to 2010 in Japan.

REFERENCES

[1] D. P. Smith, *J. Appl. Phys.* **38**, 340 (1967).

[2] M. Katayama, E. Nomura, N. Kanekuma, H. Soejima,M. Aono, *Nucl. Instr. Meth.* ***B*** **33**, 857 (1988).

[3] K. Sumitomo, K. Oura, I. Katayama, F. Shoji, T. Hanawa, *Nucl. Instrum. Meth.* ***B*** **33**, 871 (1988).

[4] M.Yoshimoto, T.Maeda,K.Shimozono, H.Koinuma, M.Shinohara, O.Ishiyama and F.Ohtani, *Appl. Phys. Lett.* ***65***, 3197 (1994).

[5] K. Umezawa, S. Nakanishi, K. Kojima, M. Yoshimura, K. Ueda, W. M. Gibson, *Phys. Rev.* ***B*** **63**, 035402 (2001).

[6] J. W. Rabalis, *Low Energy Ion Surface Interactions*, (Wiley, 1994).

Mater. Res. Soc. Symp. Proc. Vol. 1318 © 2011 Materials Research Society
DOI: 10.1557/opl.2011.919

Electron Phase Microscopy of Magnetic Fields in Ferromagnets and Superconductors

Akira Tonomura

Advanced Science Institute, RIKEN, Wako, Saitama 351-0198, Japan

Advanced Research Laboratory, Hitachi, Ltd., Hatoyama, Saitama 350-0395, Japan

Initial Research Project, Okinawa Institute of Science and Technology, Kunigami, Okinawa 904-0411, Japan

E-mail: akira.tonomura.zk@hitachi.com

ABSTRACT

Highly sensitive electron phase microscopy based on the Aharonov-Bohm (AB) effect principle has been used to observe microscopic distributions of magnetic fields in ferromagnets and superconductors. The observation examples include the unconventional behaviors of interlayer Josephson vortices in anisotropic layered high-Tc superconducting $YBa_2Cu_3O_{7-\delta}$ (YBCO) thin films, which are produced when the applied magnetic field is greatly tilted to the layer plane, and the magnetic-field distributions of tiny magnetic heads for perpendicular recording and of colossal magnetoresistance (CMR) materials.

INTRODUCTION

Electrons behave as waves as long as they are not detected. Thanks to recent developments of brighter field-emission electron beams, it is now practical to use the wave nature of electrons to directly image the quantum world by using the phase information of electrons [1]. Magnetic lines of force can be quantitatively observed as phase contours of transmitted electrons by holographic interference microscopy [2] since a phase shift of 2π is produced between two electron beams enclosing a magnetic flux of h/e due to the Aharonov-Bohm (AB) effect [3], and quantized vortices in superconducting thin films can be dynamically observed by Lorentz microscopy [4].

This paper describes the challenges we faced in our efforts to investigate the phenomena related to Josephson vortices in layered high-Tc superconductors, the magnetic characteristics of tiny magnetic heads used for high-density perpendicular recording, and the nucleation mechanism of ferromagnetic domains in CMR materials.

EXPERIMENTAL METHOD

We carried out experiments on superconductors using a 1-MV field-emission transmission electron microscope and ones on magnetic heads and CMR materials using 300-kV field-emission electron microscopes. We observed vortices in superconducting thin film as Lorentz micrographs by defocusing the electron microscope images (figure 1). The phase distribution of the transmitted electrons through magnetic flux is transformed into the corresponding intensity variations by image defocusing. When vortices penetrated a thin film parallel to the film surface as interlayer Josephson vortices, they could be observed without tilting the film (figure 1(a)).

However, when the vortices were perpendicular to the film, the film had to be tilted so that the incident electrons would interact with a perpendicular component of the vortex magnetic field (figure 1(b)).

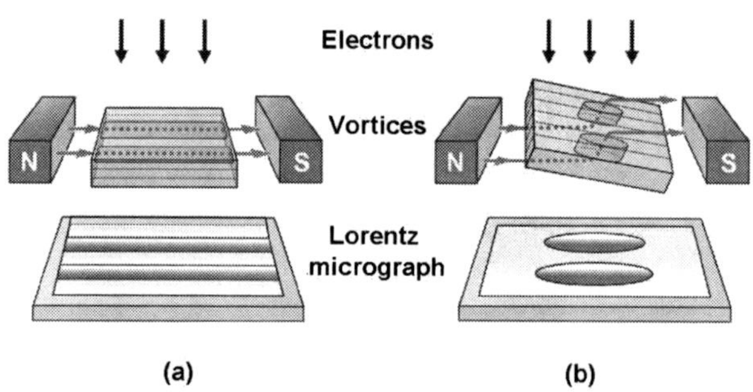

Figure 1. Experimental arrangements for observing Josephson vortices in layered superconductors by Lorentz microscopy. (a) Vortices in *ab*-plane film. (b) Vortices in *ac*-plane film. Electrons are phase-shifted by magnetic field of vortices, and the phase changes of the transmitted electrons are transformed into intensity changes by image defocusing. When vortices penetrate the film perpendicularly to it, the film is tilted as in (b). Magnetic fields in a free space like a uniform magnetic field cannot be observed by defocused Lorentz microscopy — only vortices inside superconductors, or, more exactly, their magnetic fluxes can be observed.

Projected magnetic lines of force of magnetic heads and ferromagnetic domains in CMR materials were directly observed as contour lines in the interference micrographs obtained from off-axis electron holograms formed using electron biprisms [1].

EXPERIMENTAL RESULTS

Josephson vortices in YBCO thin films

Josephson vortices in highly anisotropic $Bi_2Sr_2CaCu_2O_{8+\delta}$ (Bi-2212) superconductors not only play an important role in the magnetic characteristics of the superconductors but also are expected to be practically useful as Terahertz radiation sources. However, they had never been directly observed since they are hidden inside superconductors. Using Lorentz microscopy (figure 1 (a)) with our 1-MV microscope, we were the first to observe the static and dynamic behaviors of Josephson vortices in *ab*-plane films of moderately isotropic ($YBa_2Cu_3O_{7-\delta}$ YBCO) crystals, as shown in figure 2 [5].

Each line in the Lorentz micrograph in the figure corresponds to an individual Josephson vortex. We confirmed this by measuring a relative phase-shift of π between two electron beams passing through on both sides of a single line, which indicates that the magnetic flux of h/2e was enclosed. When the applied magnetic field strength was increased, the density of vortex lines increased proportionally with the magnetic field.

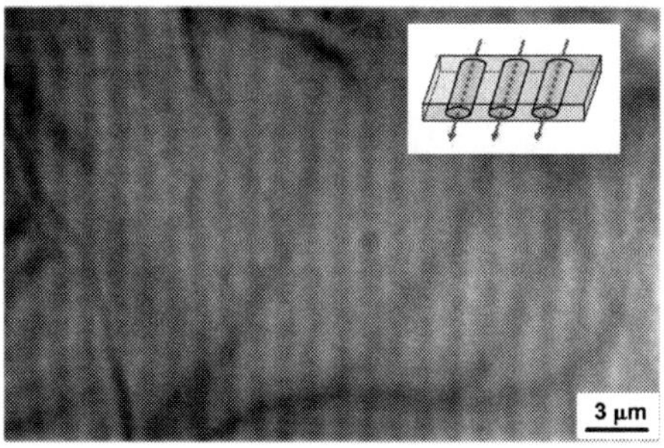

Figure 2. Josephson vortices in *ab*-plane YBCO thin film ($T = 68$ K).

The vortex lines, however, were not produced in samples thinner than 0.4 μm when the applied magnetic field was less than 100 Oe. An example Lorentz micrograph of a tapered film is shown figure 3. Two Josephson vortices are evident in the thick region on the right, while in the thin region, only the spot of a vertical vortex is evident. The contrast of the vertical vortex was due to the film being tilted around the axis determined by the direction of the applied magnetic field. It can be concluded from the bright-and-dark contrast features that the magnetic field of this vortex was upward.

Josephson vortices were produced only when the magnetic field was parallel (± 0.5°) to the layer plane under field-cooled conditions. Once produced, they behaved as if they were firmly trapped between layers. However, when the magnetic field was tilted 3°, the images of the vortex lines suddenly terminated at spots near the film edge. One end of each Josephson vortex line must have stood up perpendicularly to the film plane near the film edge due to the tilting of the applied magnetic field.

This state of the vortices was not static. The spots immediately began to move along the vortex lines, and the lines shortened and disappeared. This movement could not be stopped. This behavior of Josephson vortices indicates that they are formed by circulating elliptical vortex currents and that they cannot be easily transformed into tilted vortices consisting of disc-like vortex currents within layers.

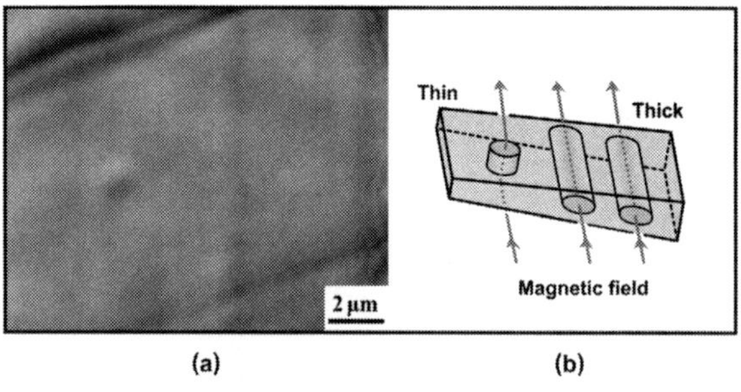

Figure 3. Vortices in tapered YBCO thin film. (a) Lorentz micrograph. (b) Schematic. Josephson vortices penetrate film thicker than 0.4 μm.

We also attempted to observe elliptical cross sections of Josephson vortices. We fabricated *ac*-plane film samples (figure 1(b)) instead of *ab*-plane film samples (figure 1(a)) and applied a magnetic field perpendicular to the film plane.

An example of a Lorentz micrograph of Josephson vortices penetrating *ac*-plane film is shown in figure 4. The images of the vortices are elongated along the *a*-axis or the *c*-plane, revealing the elliptical cross sections of the vortices. Since we are observing the vortices themselves inside superconductors, we can observe their dynamics as well. In fact, when the magnetic field was reduced, these vortices began to move only in the direction of the *a*-axis, never in different directions. This indicates strong intrinsic pinning due to the layer structure. The velocity of the vortices was 300 nm/sec on average at 5 K.

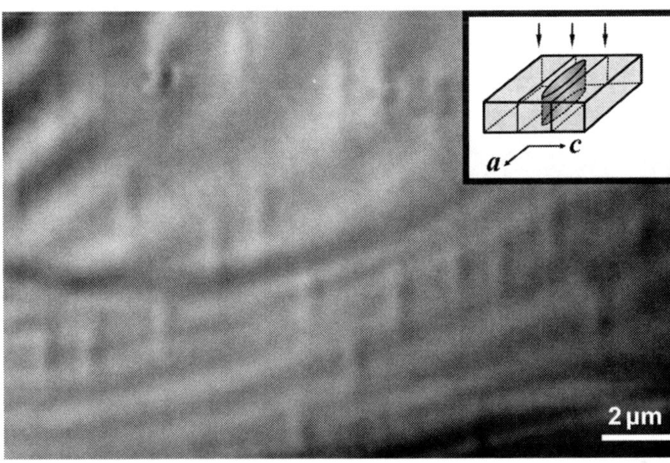

Figure 4. Lorentz micrograph of Josephson vortices penetrating perpendicularly to *ac*-plane YBCO thin film.

The elongated vortex images in the vertical direction in the micrograph show the cross sections of the vortices. The direction of the image elongation is along the *a*-axis, or within the layer plane. When the magnetic field applied perpendicularly to the film plane was switched off, these vortex images moved downward along the *a*-axis indicating the intrinsic pinning due to the layers of the YBCO crystal.

Magnetic heads

Magnetic lines of force radiating outward from a tiny magnetic head were observed as holographic interference micrographs [6] (figure 5 (a)). This head is used for high-density perpendicular recording, so it must produce a high yet localized magnetic field. A strong magnetic field is rather simple to produce. However, this is not sufficient. We also need to be able to control the field in any way we want. For example, we need to get a zero field outside a head, which poses difficulties since the density of magnetic lines inside ferromagnets is a constant determined by the strength of the magnetization.

In the case of our FeNi head shown in figure 5 (a), the straight magnetic lines inside the head were changed into "U-turn" lines [6] (figure 5 (b)) by reducing the applied magnetic field to zero. At the pole tip, which is used to localize magnetic fields, the magnetic lines formed a tiny closed circuit. The distributions in (a) and (b) of magnetic lines are symmetrical.

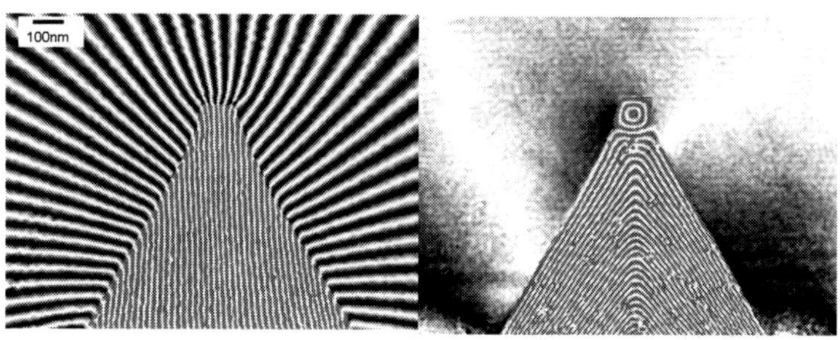

(a) (b)

Figure 5. Interference micrographs showing magnetic lines of force of an FeNi magnetic head for perpendicular recording: (a) ON; (b) OFF.
Magnetic lines are produced radially from the magnetic head when it is magnetized in the upward direction by applying an external magnetic field ((a)). When the magnetic field is removed, magnetic lines outside the head almost disappear, since magnetic lines inside it form a closed circuit ((b)). These patterns are both symmetrical. The "red" straight magnetic lines in (a) cannot be changed into the "red" U-turn and Closed-loop magnetic lines in (b) keeping the symmetry. The process of this change is shown in figure 6.

The question is how to change straight lines into U-turn lines so that the outside magnetic lines disappear. The series of Lorentz micrographs in figure 6 shows how this is done. When the applied magnetic field was reduced from 120 to 80 Oe, the straight magnetic lines inside the head (figure 6 (a)) begin to bend (figure 6 (b)). This bent region moved upward (figure 6 (c)) when the magnetic field was further reduced. The reason for this movement is evident from the pictures. The U-turn lines appeared from the lower left (figure 6 (c)), pushing the blue lines up.

The U-turn lines gradually moved upwards (figures 6 (d) and (e)) and finally covered the whole head region (figure 6 (f)). The bent magnetic lines finally changed into closed circles at the tip. The leakage magnetic field gradually decreased from (a) to (f) and almost vanished in the remanent state (f), where all the magnetic lines inside the head were closed, restoring the pattern symmetry.

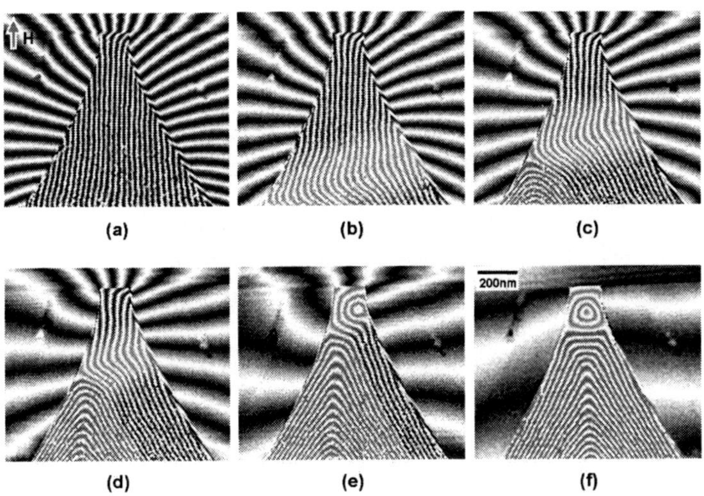

Figure 6. Interference micrographs showing how magnetic lines inside the head changed from straight lines into U-turned lines when the applied magnetic field was reduced: (a) H = 120, (b) H = 80, (c) H = 60, (d) H = 35, (e) H = 15, (f) H = 0 Oe. "Red" U-turn lines appeared from the lower left (figure 6 (c)) and moved upwards, pushing the region of "blue" bent magnetic lines (figures 6 (c), (d), (e)). The patterns of magnetic lines were no longer symmetrical. The bent magnetic lines formed a "blue" circular closed circuit in the remanent state (figure 6 (f)).

Colossal magnetoresistance

We then observed the nucleation and growth of ferromagnetic domains in colossal magnetoresistance materials, in which an electric current begins to flow when the magnetic field increases or the temperature decreases. The microscopic mechanism for this is considered to be that isolated tiny ferromagnets nucleate and coalesce in the "insulating" phase, thereby creating conduction paths of electrons. We investigated this directly using Lorentz microscopy and interference microscopy [7].

The growth of ferromagnetic domains was dynamically observed by Lorentz microscopy (figure 7 (a)) while reducing the sample temperature. The ferromagnetic domains whose magnetization is indicated by the arrows increased when the sample temperature decreased. Leakage flux from the ferromagnetic domains cannot be observed in Lorentz micrograph (a), but can be in interference micrograph (b). When we followed the growth of the ferromagnetic domains carefully by interference microscopy, it was revealed that leakage flux assisted the domain growth locally.

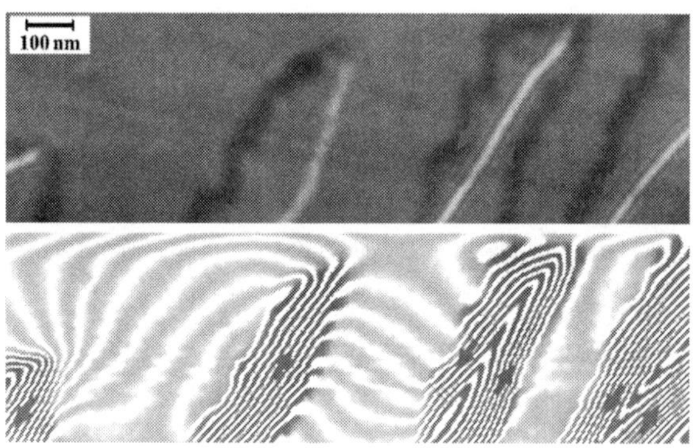

Figure 7. Growth of ferromagnetic domains in colossal magnetoresistance manganite, $La_{0.25}Pr_{0.375}Ca_{0.375}MnO_3$, thin film at $T = 62.4$ K: (a) Lorentz micrograph; (b) interference micrograph. "Blue" magnetic lines indicate magnetization in ferromagnetic domains, and yellow magnetic lines leakage fluxes from ferromagnetic domains.

RESULTS

Thanks to recent advances in technology, such as the development of coherent yet bright electron beams, the wave nature of electrons can now be used to observe microscopic objects previously unobservable by conventional electron microscopy. Specific examples include the quantitative observation of the microscopic distribution of magnetic lines in *h*/*e* units by interference microscopy and the dynamic observation of quantized vortices in superconductors by Lorentz microscopy.

Electron phase microscopy with a field-emission electron microscope is expected to play a more and more important role in future research and development not only in nanotechnology but also in biology, since microscopic information about materials reaching down even to atoms and molecules has become more pivotal.

REFERENCES

1. A. Tonomura, *Proc. Natl. Acad. Sci. USA* **102**, 14952 (2005).
2. A. Tonomura, T. Matsuda, J. Endo, T. Arii and K. Mihama, *Phys. Rev. Lett.* **44**, 1430 (1980).
3. Y. Aharonov and D. Bohm, *Phys. Rev.* **115**, 485 (1959).
4. K. Harada, T. Matsuda, J. Bonevich, M. Igarashi, S. Kondo, G. Pozzi, U. Kawabe and A. Tonomura, *Nature* **360,** 51 (5 November 1992).
5. A. Tonomura, T. Akashi, H. Kasai, S. Mamishin, T. Yoshida, T. Matsuda, S. Tajima, H. Tadatomo, T. Masui and A. Buzdin, (2009) to be submitted.
6. J. J. Kim, K. Hirata, Y. Ishida, D. Shindo, M. Takahashi and A. Tonomura, *Appl. Phys. Lett.* **92**, 162501 (2008).
7. Y. Murakami, H. Kasai, J. J. Kim, S. Mamishin, D. Shindo, S. Mori and A. Tonomura, *Nature Nanotech.* **5**, 37 (2010).

Mater. Res. Soc. Symp. Proc. Vol. 1318 © 2011 Materials Research Society
DOI: 10.1557/opl.2011.920

Effect of Oxygen Pressure on the Initial Oxidation Behavior of Cu and Cu-Au Alloys

Langli Luo[1], Yihong Kang[2], Zhenyu Liu[2], Judith C Yang[2], Guangwen Zhou[1]
[1] Department of Mechanical Engineering & Multidisciplinary Program in Materials Science and Engineering, State University of New York, Binghamton, NY 13902
[2] Department of Mechanical Engineering and Materials Science, University of Pittsburgh, Pittsburgh, PA 15261

ABSTRACT

A wide information gap exists between our present atomic-scale knowledge of metal oxidation derived from conventional ultrahigh vacuum (UHV) surface science experiments and the oxidation mechanisms obtained from the growth of bulk oxide thin films under technologically relevant realistic (or near-) atmospheric conditions. To bridge this pressure gap, we present an in-situ transmission electron microscopy (TEM) study of the initial oxidation stage of Cu(100) and Cu-Au(100) surfaces where the oxygen partial pressure varies from $5x10^{-4}$ to 150 Torr. For Cu(100), with increasing oxygen partial pressure (pO_2), the nucleation density of the oxide islands increases and so does the growth rate of the oxide islands. As the pO_2 continues to increase, a transition from epitaxial cube-on-cube Cu_2O islands to randomly oriented oxide islands is observed. A kinetic model based on the classic heterogeneous nucleation theory is developed to explain the effect of oxygen partial pressure on the oxide orientation. It is shown that such a transition in the oxide nucleation orientation is related to the effect of oxygen pressure on the nucleation barrier and atom collision rate. The Cu-Au(111) alloy revealed the same oxygen pressure dependency of the oxide nucleation orientation as pure Cu oxidation.

INTRODUCTION

Fundamental understanding of metal and alloy oxidation is of significant importance to a wide range of applications including corrosion, catalysis, fuel reactions, electrochemistry, and thin film processing. It is well known that metal oxidation involves hierarchical multiple length scales and proceeds generally from oxygen surface chemisorptions to oxide nucleation and growth and then to the formation of a continuous, macroscopically thick oxide layer. Over the past decades, surface science studies on the oxidation of metal surfaces under idealized conditions, i.e., ultrahigh vacuum (UHV), have gathered a wealth of knowledge including the surface structure, adsorption sites, and reaction pathways. However, the resulting profound understanding of the reaction mechanism obtained under rarefied UHV conditions does not automatically translate into an equally good understanding of the technologically relevant realistic (or near-) atmospheric oxidation. For example, it has been shown recently that the surface oxidation proceed readily under high pressure oxidation conditions, but appear not to proceed under the low pressure conditions typical of a vacuum experiment, despite favorable thermodynamics[1-5]. Therefore, it is paramount to bridge this pressure gap by determining the difference in the response of a metal surface exposed to low and high pressures of oxygen gas.

We have utilized in situ TEM techniques to study the initial stages of Cu and Cu-Au oxidation for bridging the information gap between the knowledge obtained from surface science methods which mainly focus on the adsorption of up to a few oxygen layers and that of traditional oxidation theories of bulk oxide growth [6-10]. In this work, we particularly examine how the difference in oxygen pressure can lead to different behaviors of oxide nucleation during the initial oxidation of Cu surfaces. We have extended our experimental system from pure Cu to Cu-Au alloys. Alloying can affect the nucleation and growth of early stage of oxide islands and then influence the microstructures of the oxide layer as well as the oxidation resistance properties. By alloying Cu with Au, the oxidation behavior can be changed both thermodynamically and kinetically in ways of altering surface and interface energies, preferred adsorption sites and diffusion paths.

EXPERIMENT

Our *in situ* oxidation experiments were carried out in a modified JEOL 200CX TEM. This microscope is operated at 200keV and equipped with a UHV chamber with a base pressure $\sim 10^{-8}$ Torr. A controlled leak valve is connected to the column of the microscope permits introduction of gases directly into the microscope. Single crystal ~1000Å of Cu and Cu-Au alloy thin films were grow on NaCl substrates epitaxially by UHV e-beam evaporator with a base pressure $\sim 10^{-9}$ Torr. The films were removed from the NaCl substrate by flotation in deionized water, washed and mounted on a specially designed TEM sample holder, which allows for resistive heating at temperatures between room temperature and 1000°C. The native Cu oxide is removed by annealing the alloy films in the TEM under vacuum conditions at ~750 °C or by *in situ* annealing in hydrogen gas at a pressure of 5×10^{-5} Torr but at a lower temperature ~350 °C resulting in a clean metal surface. Oxygen gas of 99.999% purity can be admitted into the column of the microscope through the leak valve at a partial pressure pO_2 between 5×10^{-4} and 760 Torr for 30mins. For oxidation under $pO_2 > 1\times10^{-4}$ Torr, the chamber is isolated in order to protect microscope and the *in situ* observation can be made when $pO_2 < 1\times10^{-4}$ Torr.

RESULTS AND DISCUSSION

Extensive in situ TEM oxidation studies have been carried out on Cu(100) and Cu-Au(111) surfaces to elucidate the correlation between oxygen partial pressure and the oxide growth morphologies. Fig. 1 shows the results of *in situ* bright-field TEM images of Cu(100) surfaces oxidized at 350°C and different pO_2 for 30mins. It shows that the nucleation density of Cu_2O islands increases with increasing the oxygen partial pressure from $pO_2=5\times10^{-4}$ to 50 Torr. The selected-area electron diffraction (SAED) patterns from the oxidized surfaces reveal that epitaxial Cu_2O islands (i.e., the cube-on-cube epitaxy) are nucleated under the lower pO_2 (<5Torr). It can be seen that the diffraction spots from Cu_2O islands are aligned well with that of Cu(100) surface. From pO_2=5 Torr, the pattern began to show double diffraction spots from both oxides and Cu surface since oxide islands covered most area of Cu surface. The double diffraction becomes more notable under pO_2=50Torr, where the oxide islands are coalesced to form a continuous oxide layer and the intensity of diffraction spots from continuous oxide film is stronger than that of the Cu substrate layer.

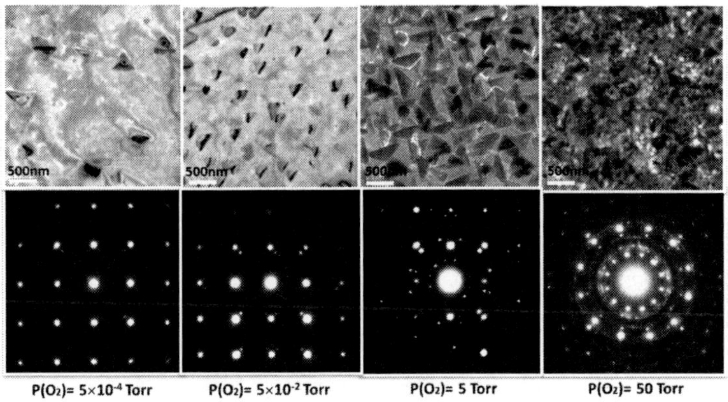

Fig.1 The bright-field TEM images and SAED patterns of Cu(100) surface oxidized at 350°C for 30mins with oxygen partial pressures from 5×10^{-4} to 50Torr

The above observation indicates that the oxide remained the epitaxial growth with the Cu substrate from the stages of nucleating oxide islands to the growth of the continuous Cu2O layer for the oxygen pressures up to 50 Torr. However, randomly oriented oxide islands are observed to nucleate for the oxidation at pO_2= 150 Torr and above. Fig. 2 shows the oxidation at the different oxygen pressures, the presence of Cu_2O diffraction rings seen in Fig.2(c) reveals the randomly orientation of the oxide islands.

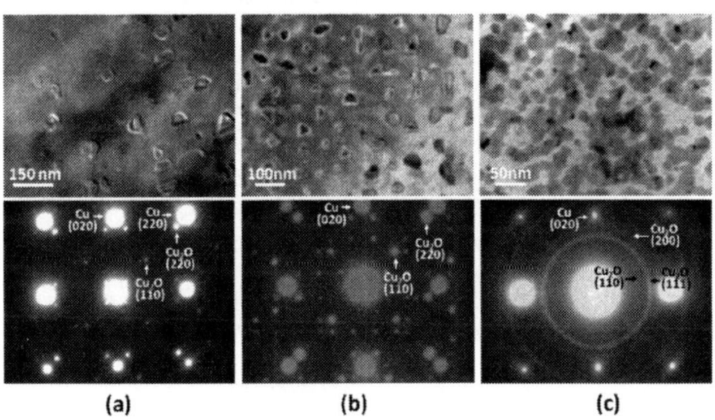

Fig. 2 (Upper panel) The bright-field TEM images of Cu_2O islands formed Cu(100) oxidized at 350°C and different oxygen pressures, (a) pO_2 = 5×10^{-4} Torr, (b) pO_2 = 0.5 Torr, and (c) pO_2 = 150 Torr; (Lower panel) SAED patterns from the corresponding oxidized Cu(100) surfaces

Figure 3 shows the high-resolution (HR) TEM images of the oxidized Cu surfaces, which further confirm this pressure-dependent orientation of oxide nuclei. Figure 3(a) is a [001] zone-axis HRTEM image obtained from the Cu surface oxidized under pO_2 = 0.5 Torr, where the two-dimensional (2D) moiré fringe pattern caused by the overlapping of Cu_2O and Cu lattices is visible. The moiré fringes running parallel to {101} lattice planes of the Cu substrate suggest the Cu_2O islands have the cube-on-cube epitaxy with the Cu lattices, i.e., the equivalent planes and directions of Cu_2O islands and the Cu substrate are matched. Figure 3(b) is a [001] zone axis HRTEM image of the Cu(100) surface oxidized under pO_2= 150 Torr, where the non epitaxial nucleation of Cu_2O islands can be evidenced by the absence of strong 2D moiré fringe contrast and the inset Fourier transform pattern of the HRTEM image.

The results indicate that the epitaxial nucleation of oxide islands cannot be maintained within the whole range of oxygen pressures. A kinetic model [11] based on the effect of oxygen pressure on the oxide nucleation barrier and atom collision rate has been developed to account for this transition of oxide nucleation orientation. The oxide nucleation rate J is obtained as

$$\frac{J}{B} = J_0 \times \exp\left(-\frac{16\pi\Omega^2(2+n)(1-n)^2}{3(kT)^3\left[\ln\left(\frac{pO_2}{pO_2^e}\right)\right]^2}\sigma_{NO}^3 \right) \quad (1)$$

with $$J_0 = \left[\frac{4(1+n)}{(2+n)(1-n)}\right]^{1/2} \times \left(\frac{pO_2}{pO_2^e}\right) \times \ln\left(\frac{pO_2}{pO_2^e}\right) \quad (2)$$

n depends on the interaction and structural match between the oxide nucleus and the metal substrate, and is related to the interfacial tension between the oxide and metal substrate by

$$n = (\sigma_{SO} - \sigma_{NS})/\sigma_{NO} \quad (3)$$

where σ_{NS} and σ_{SO} are the interfacial free energies between the oxide nucleus and metal substrate and the substrate and oxygen gas, respectively. Ω is the volume of oxygen atoms in the oxide phase; σ_{NO} denotes the specific interfacial energy between oxide nucleus and oxygen gas; k and B are constants for a certain species(e.g. Cu); pO_2 the actual oxygen pressure during the oxidation, pO_2^e the equilibrium oxygen pressure as given by the Ellingham diagram for most metal oxides[12].

The change in the nucleation orientation under different oxygen pressures can be physically understood as follows. At low oxygen pressure, the nucleation barrier is very high, and the top priority to accelerate the nucleation kinetics is to lower the nucleation barrier. Therefore, heterogeneous nucleation with the strong interaction and good structural match between an oxide island and the metal substrate will be kinetically favored. Conversely, at high oxygen pressures, the nucleation barrier is reduced and the exponential term becomes less important. Instead, the issue of effective collisions, described by the collisional pre-factor, becomes important. The nucleation of oxide islands with weak interaction and poor structural match with the metal substrate is enhanced. Kinetically speaking, in order to obtain the epitaxial oxide film on the metal substrate by oxidation, the oxygen pressure should be relatively low. If the oxygen pressure is too high, the kinetics leads to a deviation of nucleating oxide islands from the orientation of the metal substrate and the epitaxial relation will be lost.

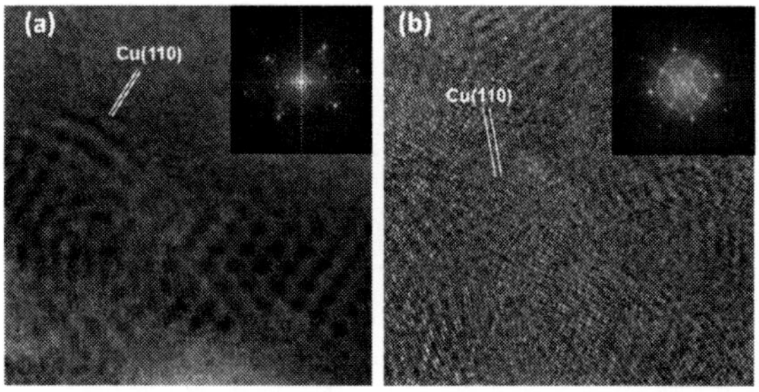

Fig. 3 HRTEM images of Cu(100) oxidized under different oxygen pressures, (a) pO_2 = 0.5 Torr and (b) pO_2 = 150 Torr. The strong 2D moiré fringe pattern implies the epitaxial nucleation of Cu_2O islands; the absence of 2D moiré fringes in (b) suggests the non epitaxial nucleation of Cu_2O islands on the surface. Inset is the Fourier transform of the HRTEM images, which reveal the similar diffraction patterns as those in Figure 2.

We then extended the various oxygen pressure experiments to Cu-Au system and examined nucleation orientations of oxide islands on Cu-Au(111) surfaces under oxygen partial pressure ranging from $5x10^{-4}$Torr to 500Torr. Fig. 4 shows the bright field TEM images and correlated SAED patterns from the oxidation under the different oxygen pressures. Similar as Cu (100), the epitaxial growth of Cu_2O islands were maintained at lower oxygen pressure (pO_2< 5Torr); the diffraction pattern starts to show deviation from the expitaxy of Cu-Au (111) surface in the medium oxygen pressure range. The selected-area electron diffraction (SAED) patterns from the oxidized surfaces reveal that randomly oriented oxide islands are nucleated for the oxidation at pO_2= 150 Torr and above, as revealed by presence of Cu_2O diffraction rings.

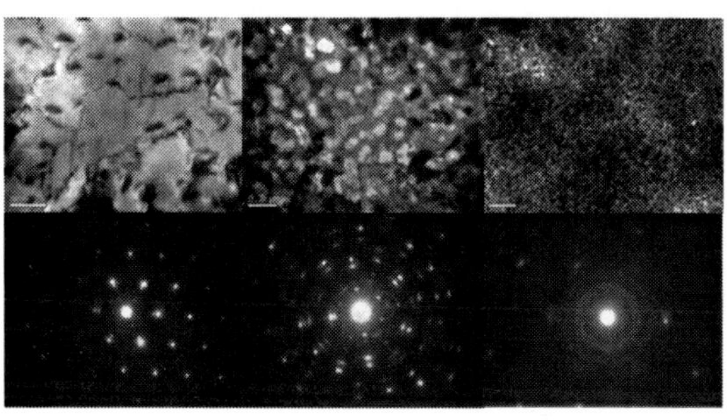

Fig. 4 Cu-Au(111) surface oxidized at 350°C for 30mins at pO_2=A) 0.005Torr; B)5Torr; C)150Torr.

CONCLUSIONS

In conclusion, the effect of oxygen pressure on the nucleation and growth of oxide islands during initial oxidation of Cu(100) and Cu-Au(111) is examined. The increase of oxygen partial pressure increases the density of oxide islands. The oxidation under low oxygen pressures leads to epitaxial oxide islands while increasing oxygen pressure results in randomly oriented oxide nuclei. It is shown that such a nucleation orientation transition is caused by the opposite effect (i.e., epitaxial vs. non epitiaxial) of oxygen pressure on the nucleation barrier and atom collision rates.

ACKNOWLEDGMENTS

This research is financially supported by the U.S. Department of Energy, Office of Basic Energy Sciences, Division of Materials Sciences and Engineering under Award No. DE-FG02-09ER46600.

REFERENCES

1. E. Lundgren, J. Gustafson, A. Mikkelsen, J. N. Andersen, A. Stierle, H. Dosch, M. Todorova, J. Rogal, K. Reuter, and M. Scheffler, Phys. Rev. Lett. **92**, 046101 (2004).
2. J. A. Eastman, P. H. Fuss, L. E. Rehn, P. M. Baldo, G. W. Zhou, D. D. Fong, and L. J. Thompson, Appl. Phys. Lett. **87**, 051914 (2005).
3. K. Lahtonen, M. Hirsimaki, M. Lampimaki, and M. Valden, The Journal of Chemical Physics **129**, 124703 (2008).
4. I. Lyubinetsky, S. Thevuthasan, D. E. McCready, and D. R. Baer, Journal of Applied Physics **94**, 7926 (2003).
5. G. W. Zhou, Phys. Rev. B **81**, 195440 (2010).
6. G. W. Zhou, J. C. Yang, Physical Review Letters **89**, 106101 (2002).
7. G. W. Zhou, J. C. Yang, Applied Surface Science **210**/3-4, 165 (2003).
8. G. W. Zhou, J. C. Yang, Surface Science **531**/3, 359 (2003).
9. G. W. Zhou, J. C. Yang, Applied Surface Science **222**, 357-364 (2004).
10. G. W. Zhou, J. C. Yang, Surface Science **559**/2-3, 100-110 (2004).
11. Langli Luo, Yihong Kang, Zhenyu Liu, Judith C Yang, Guangwen Zhou, Phys. Rev. B, Accepted on March 2nd, 2011
12. D. R. Gaskell, "*Introduction to Metallurgical Thermodynamics*" (Scripta, Washington, D.C., 1973).

Mater. Res. Soc. Symp. Proc. Vol. 1318 © 2011 Materials Research Society
DOI: 10.1557/opl.2011.921

Automated Crystallite Orientation and Phase Mapping in a Transmission Electron Microscope

Sergei Rouvimov[1,2], Peter Moeck[1], Ines Häusler[3], Wolfgang Neumann[2,3], and Stavros Nicolopoulos[4]

[1] Department of Physics, Portland State University, Portland, OR 97207-0751, U.S.A. & Oregon Nanoscience and Microtechnologies Institute, www.onami.us
[2] Department of Chemistry, University of Oregon, Eugene, OR 97401-3753, U.S.A. & Oregon Nanoscience and Microtechnologies Institute
[3] Institute of Physics, Humboldt University of Berlin, Newtonstreet 15, 12489 Berlin, Germany
[4] NanoMEGAS SPRL, Boulevard Edmond Machterns No 79, Saint Jean Molenbeek, Brussels, B-1080, Belgium, nanomegas.com

ABSTRACT

An automated technique for the mapping of nanocrystal phases and orientations in a transmission electron microscope (TEM) is briefly described. It is primarily based on the projected reciprocal lattice geometry that is extracted automatically from precession electron diffraction (PED) enhanced spot patterns. The required hardware allows for a scanning-precession movement of the primary electron beam on the crystalline sample and can be interfaced to any newer or older mid-voltage TEM. Comprehensive open-access crystallographic databases that may be used in support of this technique are mentioned.

INTRODUCTION

The electron backscattered diffraction (EBSD) technique (which is also known as orientation imaging microscopy) in a scanning electron microscope (SEM) [1] is often employed for structural characterization of polycrystalline materials. EBSD in a SEM is based on back-scattered Kikuchi diagrams. It is therefore sensitive to the plastic deformation state of the crystallites as well as to structural damage or contamination of the crystal surfaces. The spatial resolution of this technique ranges from 20 to 80 nm [1,2]. There are also ongoing efforts to develop orientation imaging microscopy in a transmission electron microscope (TEM) [2,3]. That technique is marketed as "Automated Crystallography for the TEM" (ACT) by TSL-EDAX and works on the basis of some 5,000 automatically collected conical dark field images.

In the fall of 2008, an alternative automated electron diffraction system for orientation imaging in the TEM has been demonstrated for the first time [4]. Heavily deformed Cu foils with grain sizes on the order of 20 nm were used as examples. The utilization of the PED mode [5,6] proved crucial to the success of the automated procedure on these samples. This electron diffraction technique is based on template matching of experimentally obtained electron diffraction spot patterns to their pre-calculated theoretical counterparts. Such template matching has been utilized off line for the crystallographic characterization of precipitates in heavily deformed austenitic stainless steels [7,8] and shown to be a very robust procedure.

The new technique produces crystallite orientation and phase map also for plastically deformed polycrystalline materials (which do not produce back-scattered Kikuchi lines of a sufficient quality in an SEM). Also there are no "special crystal surface" preparation steps necessary for the application of this technique. The often required crystal surface preparation methods for EBSD/SEM are simply replaced by standard TEM preparation techniques in order to obtain sufficiently thin and electron transparent crystallites.

The Brussels/Belgium based NanoMEGAS company (with AppFive of Tempe/Arizona as its American partner) offers an automated turn-key solution to PED based nanocrystal phase and orientation mapping. This paper describes this company's system very briefly and illustrates the application of the technique to a face centered cubic (FCC) metal and a semiconductor composite sample.

AUTOMATED CRYSTALLITE PHASE and ORIENTATION MAPPING from PRECESSION ELECTRON DIFFRACTION DATA

PED patterns are automatically recorded by an external digital camera while the area of interest is scanned in a step like fashion with a nanometer to tens of nanometer diameter primary electron beam. For mapping purposes, precession angles between 0.1° to 1° are typically sufficient. The data collection step is followed by semi-automatic data processing. Figure 1 gives an overview of the system.

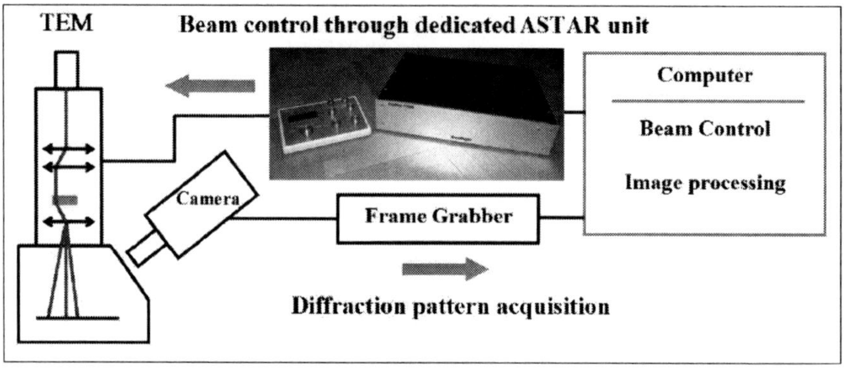

Figure 1. The experimental set up of the ASTAR/DigiSTAR system (from NanoMEGAS) for crystallite orientation and phase mapping, modified after ref. [4]. The system consists of hard- and software. The hardware can be interfaced to any older or newer mid-voltage TEM. The software runs on an external PC under Windows™ (and can also be used off line for the creation of nanocrystal orientation and phase maps from high resolution phase contrast TEM images [9]).

The orientation and phase identification is performed by the matching of each experimental diffraction pattern to a large number of templates, which are computer generated kinematical electron diffraction spot patterns for all possible orientations over the symmetry invariant section of the stereographic projection of the respective crystal point group. The degree of matching between the experimental electron diffraction spot patterns and the calculated templates is given by a correlation index. The highest value of the correlation index corresponds to the best match for the crystallite orientation and phase. A reliability index is defined as the ratio of the highest peak in the correlation index map to the second highest peak.

The software allows for the superposition of various types of maps that are all calculated on the basis of the automatically recorded data. Crystal structure information as obtainable from the comprehensive open access databases [10-14] comprises the bulk of the prior knowledge that is necessary to perform the crystallite orientation and phase mappings.

Precession of the primary electron beam around the optical axis during the recording of electron diffraction spot patterns results in more reflections with intensities that are closer to the predictions of the kinematical theory [5,6,9]. For the same electron wavelength, PED brings a larger section of the Ewald sphere in contact with the crystallite's reciprocal lattice. This is advantageous for dealing with the so called "180° ambiguity" in the indexing of electron diffracting spot patterns.

That ambiguity arises from the fact that a particular diffraction spot in the zero order Laue zone may be indexed as either (hkl) or $(\bar{h}\bar{k}\bar{l})$ [15]. While one will obtain the same zone axis indices for both sets of indexed reciprocal lattice vectors, an ambiguity will exist as to the azimuthal orientation of the crystal around the zone axis. Two azimuthal orientations that differ by 180° will be equally likely. For non-centrosymmetric cubic crystals, one may reformulate this ambiguity as being one between indexing non-symmetry equivalent zone axes as either $[uvw]$ or $[\bar{u}\bar{v}\bar{w}]$. PED spot pattern as obtained with the DigiSTAR hardware frequently contain reflections from higher order Laue zone so that this ambiguity can be reliably overcome by the ASTAR software. This is especially true for larger precession angles. Additional methods to tackle this ambiguity are discussed in ref. [16].

EXPERIMENTAL EXAMPLES

Thin polycrystalline copper films were analyzed at the FEI Tecnai G^2 F20 ST analytical field-emission microscope of Portland State University. This microscope is equipped with the ASTAR/DigiSTAR system of NanoMEGAS and serves as demo site for this company for the USA.

While the step size was set to 150 nm in these analyses in order to cover wide sample areas, a primary electron beam with a diameter of approximately 20 nm was used. (Due to the field emission source, nanometer sized beams can be utilized in this microscope together with suitably small step sizes.) A comparatively low precession angle of 0.15° was used for our analyses. The virtual bright field, Fig. 2a, crystallite orientation, Fig. 2b, correlation index, Fig. 2c, reliability index, Fig. 2d, and virtual dark field maps in the bottom row of Fig. 3 may, therefore, contain some thin film buckling artifacts. The crystallite orientation map, Fig. 2b, nevertheless shows sufficient pixel value homogeneity throughout many of the crystallites.

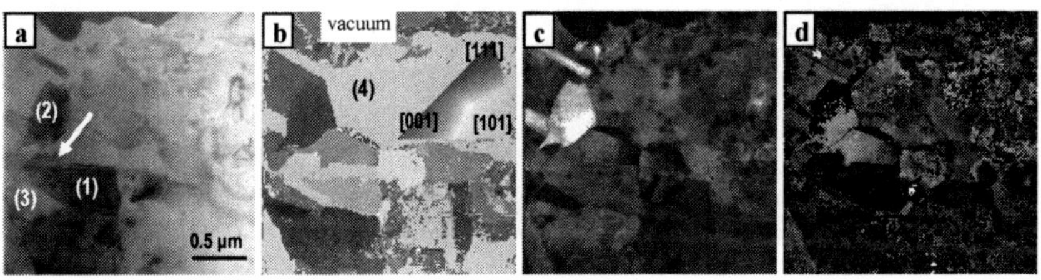

Figure 2. Crystallite orientation mapping of polycrystalline copper, 0.15° precession angle. **(a)** Virtual bright field map, **(b)** crystallite orientation map with the [001]/(red) - [101]/(green) - [111]/(blue) stereographic triangle for point group $m\bar{3}m$ (i.e. FCC crystallites) as inset, **(c)** correlation index map, and **(d)** reliability index map from the same specimen area. Note that the "microscope's vacuum area" is clearly visible at the top of both the virtual bright field map and in the dark field maps below. This area appears "light orange" in the crystallite orientation map. The numbers in round brackets in (a) and (b) identify individual crystallites that are correspondingly labeled in the virtual dark field maps in the bottom row of Fig. 3 below.

Virtual bright field images are maps of the intensity of the directly transmitted electron beam (that is "un-precessed" below the sample, ref. [5,6]) at each scanning increment. The "software equivalent" of an aperture is used to remove the diffracted beams. When the precession angle is small, the contrast in these images bears some resemblances to the contrast in classical bright field scanning transmission electron micrographs. Virtual dark field images of individual crystallites (or crystallite areas) are obtained with the ASTAR software by a click of the mouse after a suitable "software aperture" has been set around a particular reflection spot in one of the automatically recorded diffraction patterns from that crystallite. The respective virtual dark field map then shows higher intensity (brighter features) for that particular crystallite (or crystallite area) and also for all of the various sample areas that were simultaneously oriented in reflection conditions, e.g. Fig. 3 bottom row. The experimental electron diffraction spot patterns that formed the basis for the crystallite mapping process, on the other hand, can be displayed by clicking with the mouse on a crystallite in either the virtual bright field map or the crystallite orientation map. In the top row of Fig. 3, four such patterns are shown.

Due to the comparatively small precession angle, resemblances of "dynamical diffraction contrast" are present in the virtual bright field map, Fig. 2a, and the virtual dark field maps, bottom row of Fig. 3. Such a contrast feature is marked by arrows in both figures. While thickness fringes of classical dynamical diffraction contrast are a sequence of dark and bright stripes at a crystallite edge in the virtual bright field map of the marked crystallite, the contrast is reversed in the respective virtual dark field map. This kind of quasi-classical diffraction contrast at low precession angles will have its usage in more thorough off-line analyses of polycrystalline materials. Precession angles in the range of 0.5° to 1° tend to average such dynamical diffraction contrasts out (and allow, therefore, for reliable bright field tomography studies of composite crystalline materials with the DigiSTAR hardware of NanoMEGAS [17].)

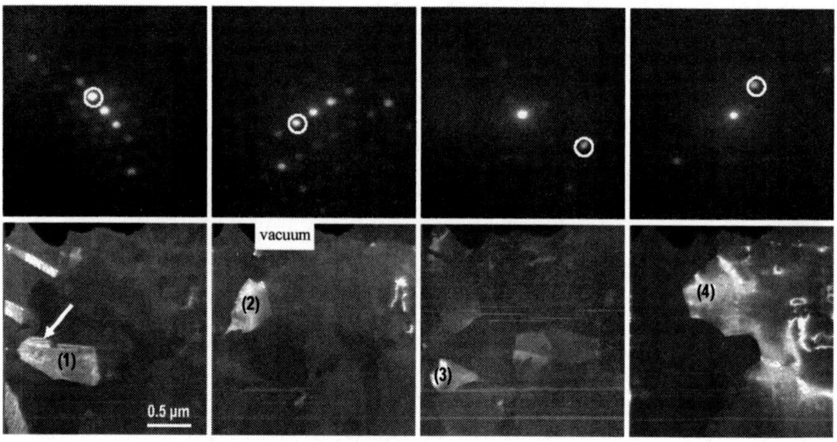

Figure 3. Crystallite orientation mapping of polycrystalline copper. **Top row:** Experimental diffraction spot patterns of selected crystallites (that appear brightest and largest in the virtual dark field maps in the bottom row). The positions of a 5 pixel diameter "aperture" that was imposed as mask on the diffraction patterns in order to create the virtual dark field maps (in the bottom row) are indicated by circles. **Bottom row:** Corresponding virtual dark field maps. The numbers in round brackets indentify individual crystallites in these maps that are correspondingly labeled in the virtual bright field map, Fig. 2a, and the crystallite orientation map, Fig. 2b, above. (The horizontal stripes in some of these maps are unimportant thresholding artifacts that are related to the after-glow effect of the microscope's viewing screen.)

At the JEOL JEM 2200FS of Humboldt University Berlin (that possesses both a field-emission gun and an in-column Ω energy filter), the data for the maps of "self assembled surface contactable partially embedded" MnAs crystallites in a GaAs matrix, Fig. 4, were collected at 200 kV. A PED device "ASTAR/DigiSTAR" from NanoMEGAS is interfaced to this microscope and it serves as demonstration site for this company for middle Europe. Both the scanning-precession step size and the probe size were on the order of magnitude 20 nm for this particular sample (but can be set to much smaller values when needed). While there was no precession during the recording of the data for the maps in Figs. 4a and 4b, a precession angle of 1° was utilized for the recording of the data in Figs. 4c and 4d.

Note that there are at least two possible candidate phases for the MnAs: the ferromagnetic α-MnAs with space group $P6_3/mmc$ and the paramagnetic β-MnAs with space group Pnma. Figure 4b shows the distribution of three identified β-MnAs crystallites (light blue) in the GaAs matrix (red) in a cross section. The cross correlation map is superimposed on this crystallite/matrix phase map. Due to the zero-precession angle, the cross correlation indices are generally low in good correspondence to the thin film buckling that is visible in the virtual bright field map of Fig. 4a.

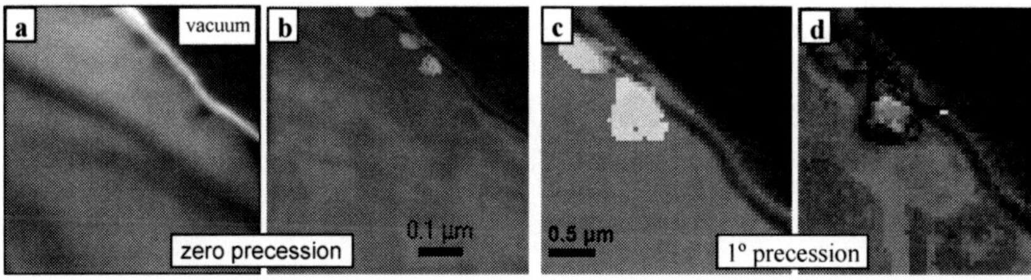

Figure 4. Crystallite phase mapping of MnAs in a GaAs matrix, [110] cross sections, to illustrate advantages of using (scanning) PED data in comparison to "on optical axis" (scanning) primary beam data. **(a)** Virtual bright field and **(b)** crystallite phase map overlaid with the corresponding correlation index map both for zero precession from the same specimen area. **(c)** Crystallite phase map overlaid with the corresponding correlation index map and **(d)** crystallite phase map overlaid with the correlation index map and the reliability index map both for a precession angle of 1° from the upper right part of the specimen area of (a) and (b). The color code in (b) to (d) is: GaAs appears in red, β-Mn in light blue, and α-Mn in green (as described in the text).

Precession at an angle of 1° enhanced the cross correlation indices everywhere, see Fig. 4c, as the thin film buckling effect was effectively removed. The cross correlation index map is also superimposed in the crystallite/matrix phase map of Fig. 4c, allowing for a direct assessment of the effect of PED data on the accuracy of the method. Note also that comparatively a larger area is mapped in Figs. 4a and 4b that is mapped in Figs. 4c and 4d. The mapped area in Figs. 4c and 4d contains only the two lower MnAs crystallites of Fig. 4b.

The distribution of two possible MnAs phases is shown in Fig. 4c in light blue for the paramagnetic orthorhombic β-phase and green for the ferromagnetic hexagonal α-phase (on the red background of the GaAs matrix). Figure 4d shows finally the superposition of the map in Fig. 4c with the corresponding phase reliability map. Since the phase reliability is low for both MnAs phases, the study of this cross section sample remained somewhat inconclusive. This may be due to the MnAs crystallites being apparently embedded in the GaAs matrix in the direction of the transmitted (primary) electron beam. Conclusive results of a study of this materials system in plan view (where the free surface of the MnAs crystallites is in the direction of the primary electron beam) will be reported together with the results of crystallite orientation and phase mappings of other materials systems in a comprehensive review [18].

COMPREHENSIVE OPEN ACCESS CRYSTALLOGRAPHIC DATABASES

As an alternative to the well known commercial databases, one may use now comprehensive open-access databases. More than 150,000 comprehensive datasets are currently available for inorganic crystals at the combined web sites of the Crystallography Open Database (COD) [10,11], the COD's mainly inorganic and educational subsets [12], the American Mineralogist Crystal Structure Database [13], and the Linus Pauling File [14].

Note that the COD has currently more than 135,000 entries and is rapidly growing [10]. All of its entries can be visualized interactively at the atomic level at one of the two Portland/Oregon based web site mirrors [10]. All datasets of the COD are in the form of Crystallographic Information Files (CIFs), i.e. the preferred format for the electronic dissemination of crystallographic data as endorsed by the International Union for Crystallography.

CONCLUSIONS

Precession enhanced automated crystallite phase and orientation mapping enables fast and reliable analyses of polycrystalline materials. Our studies suggest that this automated transmission electron microscopy based electron diffraction spot pattern technique is capable of solving problems for which there are currently no solutions by electron backscattering techniques of scanning electron microscopy.

ACKNOWLEDGMENTS

This research was supported by grants from the Oregon Nanoscience and Microtechnologies Institute. Dr. Edgar F. Rauch of the SIMAP/GPM2 laboratory, CNRS-Grenoble, is thanked for both the thin copper film sample and the special user training he provided at both Portland State University and Humboldt University Berlin. Prof. Kerstin Volz of Phillips University of Marburg is thanked for the MnAs/GaAs sample.

REFERENCES

1. D. Dingley, *J. Microscopy* **213**, 214 (2004).
2. D. J. Dingley and M. M. Nowell, *Michrochim. Acta* **147**, 157 (2004).
3. D. J. Dingley, *Microchim. Acta* **155**, 19-22 (2006).
4. E. Rauch, M. Véron, J. Portillo, D. Bultreys, Y. Maniette, and S. Nicolopoulos, *Microscopy and Analysis*, Issue 93, pp. S5-S8, November 2008.
5. R. Vincent and P. Midgley, *Ultramicroscopy* **53**, 271 (1994).
6. P. Moeck and S. Rouvimov, *Zeitschrift für Kristallographie* **225**, 110 (2010); (Special Issue on "Precession Electron Crystallography", guest editors: Chris Gilmore and Douglas Dorset).
7. E. Rauch and L. Dupuy, *Archives of Metallurgy and Materials* **50**, 87 (2005).
8. E. Rauch and M. Veron, *Mat. –wiss. u. Werkstofftech.* **36**, 552 (2005).
9. P. Moeck, S. Rouvimov, E. F. Rauch, and S. Nicolopoulos, in: *Electron Crystallography for Materials Research and Quantitative Characterization of Nanostructured Materials*, (Eds.: P. Moeck, S. Hovmöller, S. Nicolopoulos, S. Rouvimov, V. Petkov, M. Gateshki, and P. Fraundorf), Mater. Res. Soc. Symp. Proc. **1184**, 49 (2009).
10. http://cod.ibt.lt (in Lithuania), *mirrored at:* http://www.crystallography.net *and* http://cod.ensicaen.fr/ (in France), *as well as at:* http://nanocrystallography.org *and* http://nanocrystallography.net (in Oregon, USA). *Also accessible under a different search surface at:* http://fireball.phys.wvu.edu/cod/ (in West Virginia, USA). *Of these six web sites,* http://nanocrystallography.net *is currently the only mirror where the atomic structures can be visualized interactively.*
11. S. Gražulis, D. Chateigner, R. T. Downs, A. F. T. Yokochi, M. Quirós, L. Lutterotti, E. Manakova, J. Butkus, P. Moeck, and A. Le Bail, *J. Appl. Cryst.* **42**, 726 (2009); open access: http://journals.iucr.org/j/issues/2009/04/00/kk5039/kk5039.pdf.
12. http://nanocrystallography.research.pdx.edu/CIF-searchable/cod.php, data on some 20,000 crystals.
13. http://rruff.geo.arizona.edu/AMS/amcsd.php, data on some 10,000 minerals.
14. http://crystdb.nims.go.jp, data on some 30,000 metals and alloys.
15. P. E. Champness, *Electron Diffraction in the Transmission Electron Microscope,* Oxford, BIOS Scientific Publishers Ltd., (Royal Microscopical Society MICROSCOPY HANDBOOKS, vol. 47), 2001, pp. 35-36.
16. E. F. Rauch and L. Dupuy, *J. Appl. Cryst.* **39**, 104 (2006).
17. J. M. Rebled, L. Yedra, J. Portillo, S. Estradé, and F. Peiró, *Proc. IMC 17*, Rio de Janeiro, 2010.
18. P. Moeck, S. Rouvimov, E. F. Rauch, M. Véron, H. Kirmse, I. Häusler, W. Neumann, D. Bultreys, Y. Maniette, and S. Nicolopoulos, *Cryst. Res. Technol.*, accepted for the 2011 special issue on "Novel Developments in Electron Diffraction".

In Situ X-Ray Synchrotron Radiation Spectroscopies in Energy-Related Materials Science and Heterogeneous Catalysis

Mater. Res. Soc. Symp. Proc. Vol. 1318 © 2011 Materials Research Society
DOI: 10.1557/opl.2011.437

Electronic Structures of Non-Pt Carbon Alloy Catalysts for Polymer Electrolyte Membrane Fuel Cells Revealed by Synchrotron Radiation Analyses

Masaharu Oshima[1, 2], Hideharu Niwa[1], Makoto Saito[1], Masaki. Kobayashi[1, 2], Koji Horiba[1, 2], Yoshihisa Harada[1, 2], Kiyoyuki Terakura[3], Takashi Ikeda[4] , Jun-ichi Ozaki[5], Yuta Nabae[6] and Seizo Miyata[6]
[1] School of Engineering, The University of Tokyo, 7-3-1 Hongo, Bunkyo-ku, Tokyo 113-8656, Japan
[2] Synchrotron Radiation Research Organization, The University of Tokyo, 7-3-1 Bunkyo-ku, Tokyo 113-8656, Japan
[3] Research Center for Integrated Science, Japan Advanced Institute of Science Technology (JAIST), 1-1 Asahidai, Nomi Ishikawa 923-1292, Japan
[4] Quantum Beam Science Directorate, Japan Atomic Energy Agency (JAEA), SPring-8, 1-1-1 Kouto, Sayo-cho, Sayo-gun, Hyogo 679-5148, Japan
[5] Department of Chemical & Environmental Engineering, Graduate School of Engineering, Gunma University, 1-5-1 Tenjin-cho, Kiryu, Gunma 376-8515, Japan
[6] Department of Organic and Polymeric Materials, Graduate School of Science and Engineering, Tokyo Institute of Technology, 2 Ookayama, Meguro-ku, Tokyo 152-8552, Japan

ABSTRACT

The oxygen reduction reaction (ORR) mechanism of non-Pt catalysts was elucidated by investigating electronic structures of carbon alloy catalysts (CACs) for polymer electrolyte membrane fuel cells (PEMFC). For metal phthalocyanine-based carbon alloy catalysts with 1-2% of nitrogen and less than 0.1% of Co or Fe, photoelectron spectroscopy, X-ray absorption spectroscopy (XAS), and X-ray Absorption Fine Structure (XAFS) have revealed that Co or Fe metal sites are not the ORR active site, whereas the carbon atoms adjacent to graphitic nitrogens at a zigzag edge of graphite may act as the ORR active site, which is in good agreement with first principles calculations. These studies have also revealed that the zigzag edges, detected as a XAS shoulder component, are well evolved for catalysts pyrolyzed at 600°C, which show the maximum ORR activity. Based on these analyses, we have realized significantly improved carbon alloy based ORR performance, up to about 70 % of that with Pt catalysts. Further, to analyze the electronic structure of CACs during operation, we constructed a new *in situ* soft X-ray emission spectroscopy system with very high energy resolution (ΔE) of 120 meV at 640 eV at the University of Tokyo beamline BL07LSU in SPring-8.

INTRODUCTION

In the attempt to realize a low carbon emission society, various efforts have been devoted to the development of low energy consumption devices such as extremely low power large scale integration (LSI) devices and non-volatile memory storage, power generation devices such as solar cells and fuel cells, and energy storage devices such as Li ion rechargeable batteries. Among these devices fuel cells are regarded as the most promising for power generation because of their high energy-conversion efficiency. PEMFCs in particular have attracted great attention and have been implemented in vehicular and stationary applications for the conversion and storage of energy. However, because of the cost and scarcity of platinum and the low durability of platinum cathodes, several issues need to be resolved in order for PEMFCs to become commercially viable. The development of alternative non-Pt ORR catalysts is therefore of great interest for use in PEMFCs.

Among the possible replacements for Pt, carbon-based cathode catalysts such as carbon alloy catalysts (CACs) have been extensively studied because of their high ORR catalytic activities [1-8]. Carbon alloys are carbon materials doped with other elements such as boron, nitrogen, and other transition metals (TMs). Because these catalysts can be synthesized from a wide variety of materials such as organometallic complexes, iron or cobalt salt, resin, carbon black, and ammonia, the resources required for preparation of non-platinum ORR catalysts are abundant. Notwithstanding, the mechanism of the catalytic activity of carbon-based cathode catalysts is not yet well understood, even though it is recognized that an understanding of the origin of the ORR activity in carbon-based cathode catalysts is essential for the design of catalysts that can surpass the ability of conventional Pt catalysts in terms of open circuit voltage and durability. Thus far, numerous ORR mechanisms and models for carbon-based cathode catalysts have been proposed, and these can be classified into two main groups. The first is the TM-related model where TM-N_4/C, TM-N_2/C, and/or TM-C-N moieties (TM = Co, Fe) act as the ORR active site. The other is the nitrogen-related model in which nitrogen plays a crucial role in the ORR of carbon-based cathode catalysts. In the latter, the edge of the carbon layer in CACs is regarded as the ORR active site.

In order to determine the ORR active site in carbon-based cathode catalysts, a fundamental understanding of the electronic structures of CACs is required. In this study, we analyzed the electronic structures of carbon alloy catalysts using various synchrotron radiation techniques such as hard X-ray and soft X-ray photoelectron spectroscopy, XAS, and XAFS. Furthermore these electronic structures were compared with first-principles calculations and electrochemical measurements to elucidate the ORR mechanism and to facilitate further improvements to the ORR activities.

EXPERIMENT

Co-related CACs and Fe-related CACs were synthesized as follows: Co-related CAC was prepared by pyrolysis of mixtures of cobalt phthalocyanine and phenolic resin (CoPc/PhRs) at temperatures of 800°C (Co800), 900°C (Co900), and 1000°C (Co1000) followed by ball milling. The amount of Co in the mixture of CoPc/PhRs was 3 wt.%. After ball milling, the sample (e.g., Co1000) was washed with hydrochloric acid (HCl) to remove cobalt, and named (e.g., Co1000aw) [3]. For comparison, Nitrogen-doped carbon alloys were also prepared by an ammo-oxidation method, where nitrogen is doped into carbon black (Ketjen Black) with 50% ammonia gas (AO50) at 600°C.

For Fe-related CACs, the catalysts were prepared by pyrolyzing mixtures of Fe phthalocyanine and phenolic resin (FePc/PhRs). The amount of Fe in the mixture was also 3 wt.%. The FePc/PhRs was pyrolyzed under a flow of inert gas at various temperatures of 200°C, 400°C, 500°C, 550°C, 600°C, 650°C, 700°C, and 800°C for 5 h. A few samples of the catalysts for the electrochemical measurements were washed with HCl to remove Fe species. The catalyst powder was stirred in conc. HCl, and then filtered [8].

The catalytic activities of the samples were tested by linear sweep voltammetry (LSV, 1 mV/s) with a rotating disk electrode (RDE). The catalyst powder was dispersed in a mixture of Nafion solution (5 wt.%), ethanol, and water and applied onto the surface of a glassy carbon electrode. The amount of loaded catalyst was 0.2 mg/cm^2. The rotation speed of the RDE was kept at 1500 rpm, and electrochemical data were collected with a potentiostat. LSV was first carried out in a N_2-saturated H_2SO_4 solution (0.5 M), and then in an O_2-saturated H_2SO_4 solution. The ORR current was determined by subtracting the N_2 current from the O_2 current.

Synchrotron radiation analyses were performed using hard X-ray photoelectron spectroscopy (HXPES), soft X-ray photoelectron spectroscopy, XAS, XAFS, and X-ray emission spectroscopy (XES). Figure 1 shows the measurement principles of PES, XAS, and XES. The probing depth of HXPES is more than twice the inelastic mean free paths (IMFPs) of photoelectrons, that is approximately 11 nm for the N 1*s* core level and 10 nm for the Co 2*p* and Fe 2*p* core levels, respectively, when the incident photon energy is 5.95 keV. HXPES measurements carried out at BL15XU (hν=5.95 keV) and BL47XU (hν=7.94 keV) of SPring-8 were conducted in an ultrahigh-vacuum chamber (base pressure $\sim 3\times 10^{-7}$ Pa) at room temperature. The overall energy resolution at both beamlines was approximately 250 meV. The binding energies were calibrated using the Au 4*f* line from a fresh Au sample. The detection angle of photoelectrons was set at approximately 1° from the surface normal to gain detection efficiency. For a detailed chemical analysis, the backgrounds of the core level spectra were subtracted by the Shirley method. Each spectrum was fitted with Voigt (Gaussian and Lorentzian) functions, where the Gaussian width was an adjustable parameter, and the Lorentzian width for the N 1*s* was fixed to 250 meV. Soft X-ray photoelectron spectroscopy was also performed at BL27SU of SPring-8 to obtain surface sensitive information on TM nano-particles inside CACs, using the incident photon energy of 1080 eV, with the IMFP of 1.1 nm. XAFS measurements were was carried out at BL19XU of SPring-8 in order to clarify the local structures of Fe and Co in the CACs. Furthermore, XAS measurements were conducted at BL27SU of SPring-8 and at BL07A of KEK-PF to reveal the edge states of graphite-like structures in the CACs by C 1s XAS and the change in N-related precursors during heat treatment by N 1s XAS.

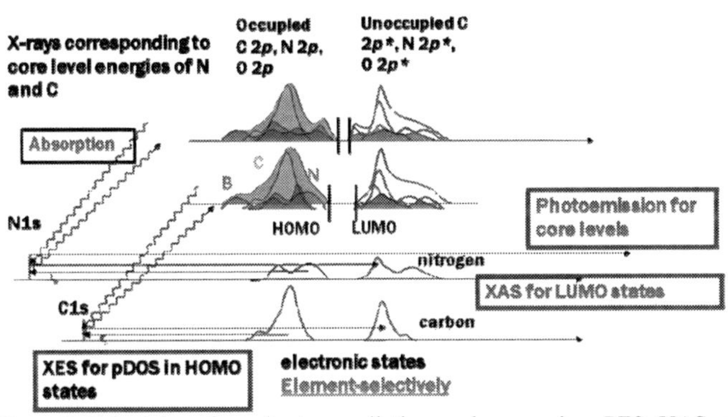

Figure 1. Principles of synchrotron radiation analyses such as PES, XAS, and XES.

RESULTS AND DISCUSSION

Chemical Structures of CACs by HXPES and XAFS

As for the crystal structure of CACs, Transmission Electron Microscope (TEM) and X-ray Diffraction (XRD) analyses indicated partial crystallization or graphitization as the pyrolyzing temperature was increased. The ORR activities were improved during the graphitization or nanoshell framework formation. Energy Dispersive X-ray spectroscopy (EDX) analysis in TEM also reveals that Fe nanoparticles with diameter of about 10 nm, shown as the dark image, coexist inside the nanoshell framework, as shown in Fig. 2(a). TEM observation revealed that the HCl acid washing effectively removed TM nanoparticles from the inside of the framework, as shown in Fig. 2(b). Therefore, it is postulated that the graphene or graphite-like framework grows by the Yarmulke mechanism [9] with Fe nanoparticles acting as a catalyst in a manner similar to carbon nanotubes (CNT). Because the ORR activities of FePc/PhRs are superior to the activities of the CACs derived from H_2-phthalocyanine and phenolic resin (H_2Pc/PhRs), it is thought that the TMs play an important role in ORR. However, it is not clear whether TM itself is the ORR active site, or if the TM acts as a catalyst in the formation of the graphite-like structure, including the ORR active site.

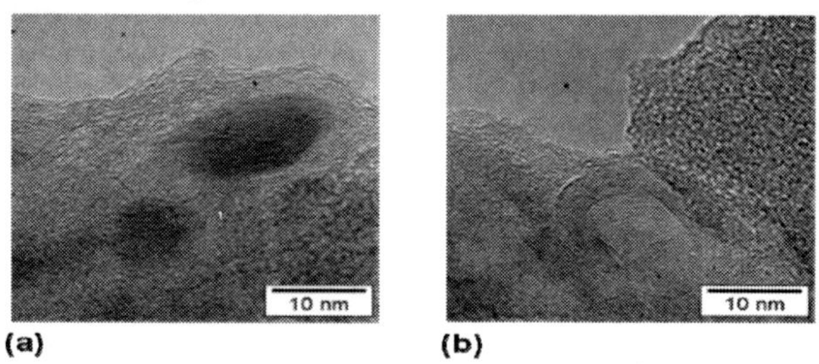

Figure 2. TEM images of FePc/PhRs CACs pyrolyzed at 600°C. (a) before and (b) after acid washing.

Synchrotron radiation analysis was performed on various CACs with different ORR activities to reveal the ORR mechanism through active site determination. HXPES has the merit of probing deeper regions than conventional XPS with better energy resolution of about 250 meV. HXPES analysis revealed that almost all the CACs contained about 1% nitrogen and less than 1% Fe or Co. Co and Fe concentrations in Co1000 and Fe800 were determined to be 0.05 at.% and 0.13 at.%, respectively, whereas the concentrations of Co and Fe after acid washing were 0.03 at.% and 0.05 at.%, respectively. Figure 3 shows the Co 2*p* core level photoelectron spectra before and after acid washing. It is clearly observed that Co^{2+} exists along with Co metallic states, and that more than half of the Co content is removed by the acid washing. A similar result was also observed also for the FePc/PhRs CACs. However, the catalytic activity of both species before and after acid washing remains almost unchanged with reduction potentials of 0.75 V at -10 mA/cm^2 for CoPc/PhRs and 0.77 V at -10 mA/cm^2 for FePc/PhRs. Therefore, it can be concluded that Co and Fe themselves are not the ORR active sites.

The pyrolysis temperature dependence of Fe chemical states in FePc/PhRs CACs was investigated using HXPES and extended X-ray absorption fine structure (EXAFS). As shown in Fig. 4, the oxidation state of Fe is reduced when the pyrolysis temperature is increased to 800°C.

EXAFS analysis clearly reveals that the number of Fe-N and Fe-O bonds drastically diminish at 800°C, with concomitant formation of Fe-Fe bonds. These observations suggest that Fe nanoparticles exist in the metallic state inside the carbon nanoshell framework, which coincides with TEM and EDX observations. Since it is not possible to completely eliminate transition metal components in CACs synthesized from TM-phthalocyanine precursors by high-temperature pyrolysis and acid washing, as shown by HXPES and EXAFS, we could not totally exclude the possibility that the TMs are ORR active sites.

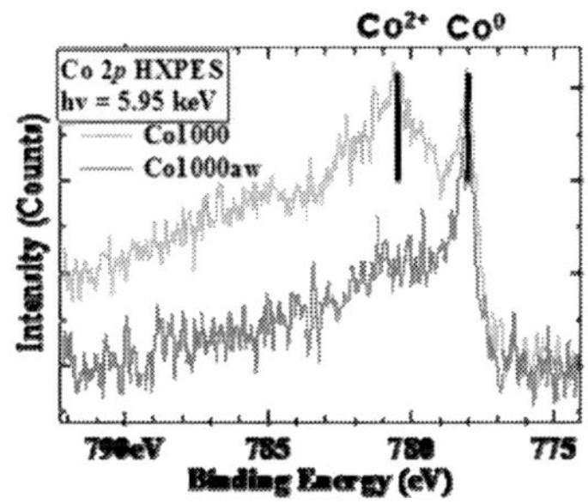

Figure 3. HXPES Co $2p$ core level spectra in CoPc/PhRs CACs before and after acid washing.

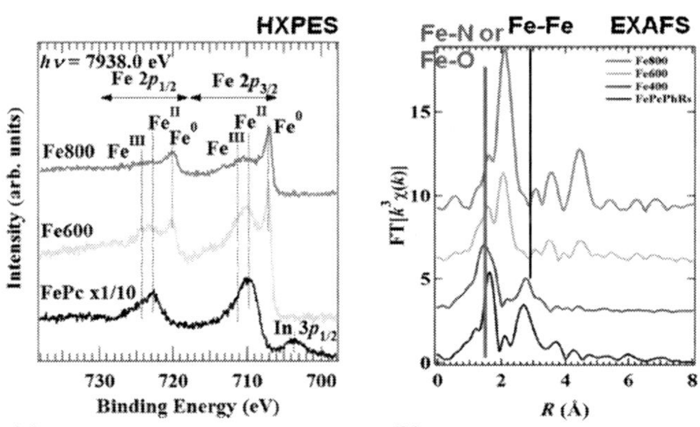

Figure 4. HXPES Fe $2p$ spectra (a) and EXAFS radial distribution function (b) for FePc and its pyrolyzed catalysts at various temperatures.

The role of impurity nitrogen in the nanoshell structure of stacked nanographene was investigated using photoelectron spectroscopy. It is well known that graphene has no density-of-states (DOS) at the Fermi level at the so-called Dirac point [10, 11]. Therefore, oxygen molecules would not adsorb on the graphene plane, resulting in no oxygen reduction reaction. However, as shown in the TEM image in Fig. 2, CACs with curved structures have many edges or defects that may produce ORR active sites. Ikeda *et al.* [12] simulated the free energies of oxygen molecule adsorption on nanographene stack with and without nitrogen doping, and found that zigzag edges are active in O_2 adsorption. Further, they revealed that the most active site for O_2 adsorption is the zigzag edge carbon with N at the position ("edge-1 N") next to the edge C atom is. On the other hand, N doping or substitution at the edge C atom, resulting in the formation of so-called pyridinic nitrogen, shows almost no activity for adsorption of oxygen molecules. Furthermore, C atoms with N next to C at the armchair edge are not active sites at all. Ikeda *et al.* also performed molecular dynamics simulation for the adsorption of O_2 and the desorption of water molecules. These first-principles simulations suggest that CACs possess potential ORR catalytic activity that is enhanced by N atoms doped at specific sites. An O_2 molecule is preferentially adsorbed associatively at C sites on graphene-like zigzag edges if a graphite-like N is located nearby.

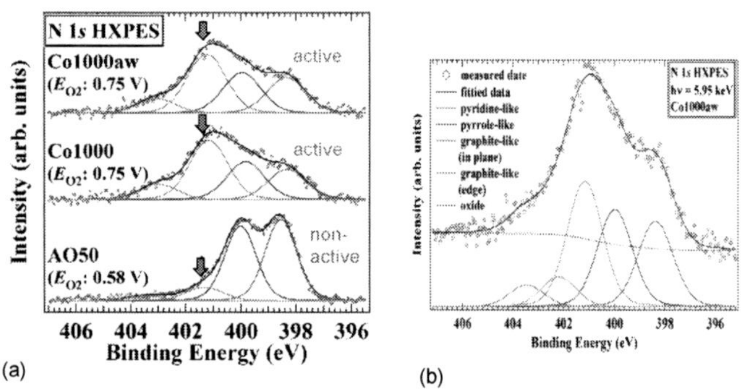

Figure 5. N 1*s* core level photoemission spectra for various CACs (a) and deconvoluted spectra for Co1000aw (b).

N 1*s* photoelectron spectra have revealed that there is a tendency that CACs with more graphitic N and less pyridinic N show higher ORR activity, whereas CACs with more pyridinic N and less graphitic N exhibit lower ORR activity [13]. It should be noted, however, that the first-principles calculation also predicts that graphitic (quaternary) N in the basal plane is not an active site at all. Therefore, it is important to deconvolute the graphitic N 1*s* peak into the basal plane component and the zigzag edge component. Figure 5 shows the N 1*s* photoelectron spectra for representative CACs showing poor ORR activity along with the spectra of the most active CACs. Given that the ratio of graphitic nitrogen is about 40% of the total N concentration (about 0.5 at.%) in this CAC pyrolyzed at 1000°C, the concentration of graphitic nitrogen becomes 0.2 at. %. Casanovas *et al.* [14] calculated the N 1*s* photoelectron spectra of quaternary N from honeycomb-like organic molecules using the first-principles calculation based on density functional theory, showing that the N 1*s* peak for the zigzag edge N component is located at 1.0 eV higher binding energy than that at the basal plane. They also found that as the size or number of honeycomb-like organic molecules increased, the chemical shift difference between the two types of N atoms decreased. The deconvoluted graphitic N 1*s* peaks are shown in Fig. 5 (b) for Co1000aw, which is the most active CAC. The ratio of the zigzag edge component to the basal plane component is about 1:4. Therefore, the concentration of the zigzag edge N is calculated to be 0.04 at.%, which is 20% of the concentration of graphitic nitrogen (0.2 at.%). Compared with the nitrogen concentration, the concentration of Co in Co1000aw is only 0.03 at. % within the probing depth, which is sufficiently deep to cover the entire area of the nanoshell. As shown in Fig. 3, almost all Co atoms are in the metallic state, which reportedly exhibits no ORR activity [15]. Therefore, these results indicate that the residual Co atoms do not play an important role in ORR contrary to other reports on Co-containing ORR catalysts.

The N 1*s* spectra of Co1000 and Co1000aw are quite similar, indicating that the acid washing did not change the chemical state of the nitrogen. It is notable that for the highly active ORR catalysts such as nanoshell carbons (Co1000 and Co1000aw), the relative composition ratio of graphitic nitrogen is the largest among the four nitrogen components. In other words, nanoshell carbons have a higher graphite-like nitrogen component than other nitrogen components. These results are consistent with our previous XAS study [16]. It should be noted that the nanoshell carbons contain less nitrogen than the other catalysts such as AO50 and AO90 that have lower ORR activities, indicating that the total amount of nitrogen is not necessarily proportional to the ORR activity of the CACs. Thus, by selectively increasing the concentration of graphite-like nitrogen at the zigzag edge, carbon alloy catalysts have the potential to become one of the most promising cathode catalysts made without noble metals.

Electronic structure of CACs by XAS and HXPES

In order to elucidate the ORR mechanism, it is important to understand the electronic structure of CACs. Huang *et al.* [17] calculated the HOMO and LUMO structures of nanographene with N and B doping using the first-principles calculations based on density functional theory. As shown in Fig. 6, it was found that the carbon atom at the zigzag edge in N doped samples has a DOS for up spin electron and another DOS for down spin electron less than 200 meV below the Fermi level, indicative of a fully filled edge state, whereas in the absence of N doping, the carbon atom at the zigzag edge has a DOS for the up spin electron only, that is, a half-filled edge state. These fully filled edge states may be responsible for the ORR active site.

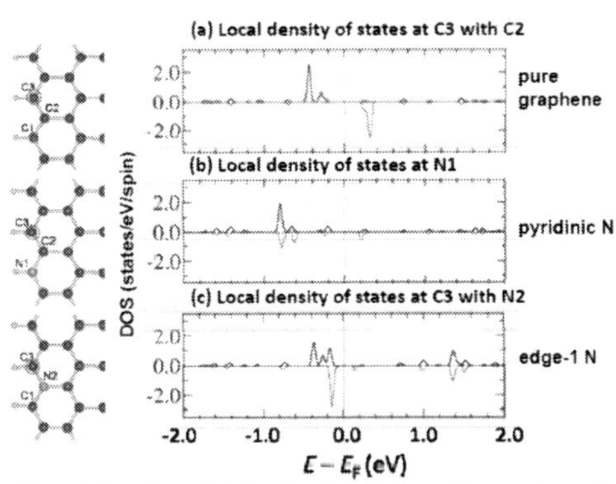

Figure 6. Calculated local density of states for (a) zigzag edge carbon atom of pure graphene, (b) pyridinic N and (c) zigzag edge carbon atom adjacent to doped nitrogen [17].

Given that the zigzag edge N concentration is less than 0.1 at.%, the ratio of the fully-filled DOS to the half-filled DOS should be quite small. It is estimated to be of the order of several percentage points. Therefore, we tried to detect the zigzag edge state using the X-ray absorption technique. Entani *et al.* [18] detected zigzag edge states in nanographene grown on Pt(111) surface by C 1*s* XAS. Joseph Joly *et al.* [19] also detected the zigzag edge states (E*) as a shoulder peak of 285 eV for the dominant π* peak in nanographene with different sizes by C 1*s* XAS and found that the peak intensity ratio of E*/π* decreases with increasing size of the nanographene formed at higher CVD temperature. Figure 7 shows the C 1*s* XAS spectra for the CACs from phenolic resin and FePc [20] pyrolyzed at 400°C, 600°C and 800°C. It is observed among the three CACs, that the shoulder peak (A) located at 284 eV is largest for the 600°C FePc sample. However, precise first-principles calculations reveal that although this shoulder peak (A) at 284 eV is not directly attributed to the zigzag edge state, it is definitely related to the zigzag edge carbons [21]. Therefore, the peak ratio of A/π* in C 1*s* XAS can be used as an indicator of the zigzag edge portion.

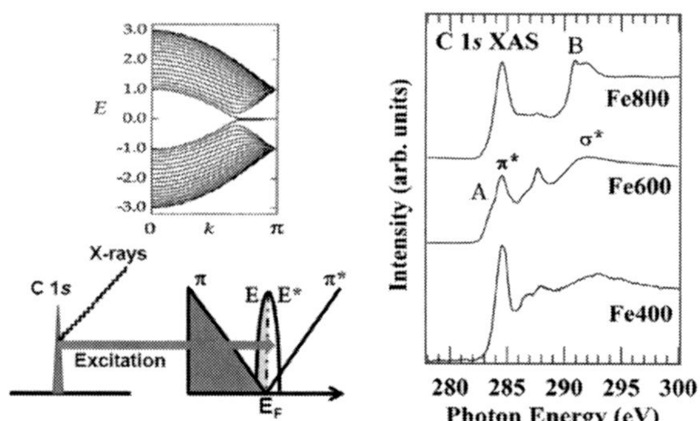

(a) (b)

Figure 7. C 1*s* XAS spectra for CACs fabricated at 400°C, 600°C and 800°C from Fe-phthalocyanine and phenolic resin (FePc/PhRs).

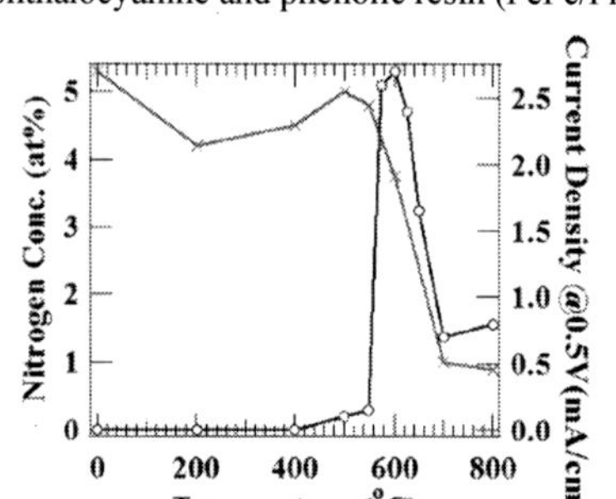

Figure 8. ORR activities and nitrogen concentration in FePc/PhRs CACs as a function of pyrolysis temperature [8].

It is found that the CACs with the larger peak ratio of A/π* in C 1*s* XAS show the higher observed ORR activities that are shown in Fig. 8. Nabae *et al.* [8] reported that this unique ORR feature may be attributed to the increase in electrical conductivity as a result of the formation of the graphite or graphene framework up to 600°C and the desorption of N above 600°C. The appearance of the zigzag edge state coincides with the growth of the graphene framework. We also observed the excitonic peak (B) at 292 eV of photon energy in the σ* region for the 800°C sample only. Because the excitonic peak is characteristic of graphite, the decrease in the intensity of the shoulder is thought to correspond to an increase in the size of the graphite-like structure, resulting in a decrease in the zigzag edge portion. Therefore, this may represent the optimum pyrolysis condition where adequate zigzag edge is formed with a considerable amount of remaining N to supply the platform for ORR active sites.

These results suggest a potential strategy for the improvement of ORR performance, by increasing the ORR active sites via increasing the zigzag edge density itself and the graphitic nitrogen concentration at the zigzag edge. In general, fuel cell current I_k can be expressed by the following equation,

$$I_k\ (\mathrm{A\ g^{-1}}) = \mathrm{SD\ (site\ g^{-1})} * \mathrm{ATF\ (e^{-1}\ site^{-1}\ s^{-1})} * e\ (\mathrm{C = A \cdot s}) \qquad (1)$$

where SD is the site density and ATF is the average turnover frequency. Here, the ATF is proportional to the electrical conductivity and reaction rate at the active site. Given that Nabae *et al.* [8] reported that electrical conductivity increases with increasing pyrolysis temperature, the improvement in the ORR activity can be attributed partly to this electrical conductivity increase in the cathode. Thus, tough or robust graphite-like structures with adequate size to preserve zigzag edge density and electrical conductivity are needed to improve both the ORR ability and the durability of PEMFCs. Wu *et al.* [22] reported the improved performance and the prolonged durability of CACs that were prepared by the acid washing and successive heat treatments of pyrolyzed FePc/PhRs up to 1000°C to preserve the zigzag edge and nitrogen concentration and strengthen graphite-like structures. The power density reached about 70 % of Pt/C catalysts, as shown in Fig. 9.

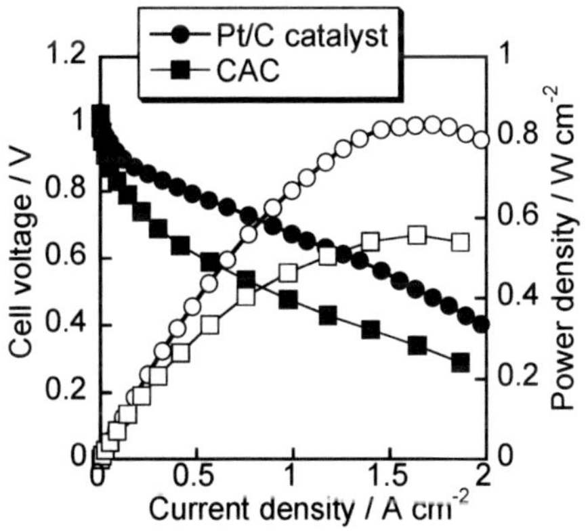

Figure 9. Fuel cell performance with a CAC improved by optimization of pyrolysis and post treatment conditions. Cathode: CAC 4 mg cm^{-2} or Pt/C 0.5 mg-Pt cm^{-2}, O_2 0.2 MPa. Anode: Pt/C catalyst 0.5 mg-Pt cm^{-2}, H_2 0.2 MPa. Electrolyte: Nafion-115. *T*: 80°C [22].

In order to observe the DOS of the ORR active site just below the Fermi level directly, valence band spectra of various CACs were measured using HXPES. As shown in Fig. 10, the valence band consists of mainly carbon $2p$ and $2s$ states together with a small amount of O $2p$ and $2s$ states from adsorbed oxygen. Although there is a tendency for CoPc/PhRs and FePc/PhRs-derived CACs with better ORR activities to show slightly larger density-of-states just below the Fermi level, as indicated by the arrow in Fig. 6, the zigzag edge states that should be located just below E_F are not clearly observed in the HXPES spectra. This may be because the ratio of fully-filled edge states to both the graphene π state and the C sp^2 state (~0.1 %) is too small, and probably because electrons in the zigzag edge states are transferred to the $2p$ state of the adsorbed oxygen. Although the C 1s XAS shoulder peak of A suggests that a considerable amount of the half-filled zigzag edge state exists, the adsorption of oxygen or OH species may also reduce the edge state in the valence band spectra.

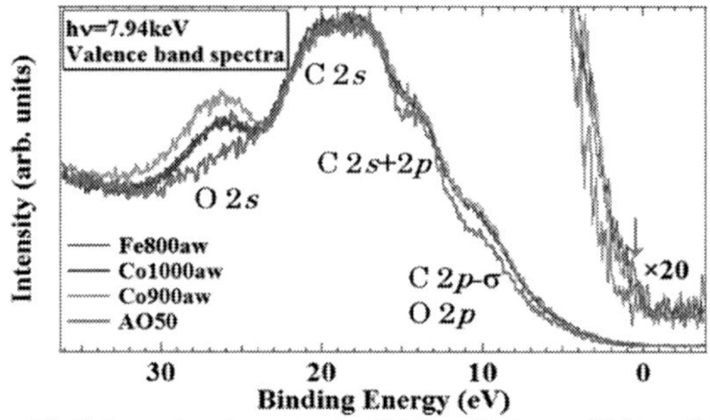

Figure 10. Valence band spectra for CACs with better ORR activities (CoPc/PhRs and FePc/PhRs-derived CACs) and CAC with poor ORR activity (AO50) measured by HXPES.

So far, there have been various studies to detect zigzag edges of graphene or graphite using scanning tunneling spectroscopy (STM) and angle-resolved photoelectron spectroscopy. Kobayashi *et al.* [23] observed that the STS peak located at 50 meV below the Fermi level increases at the zigzag edge. Sugawara *et al.* [24] observed a weak dispersion in the band diagram (E-K mapping) around 100 meV below the Fermi level along the Γ-K-M direction using ARPES for a cleaved kish graphite sample. It is suggested that further investigation of zigzag edge states should be done using in-vacuum annealing of CACs.

In situ **electronic structure analysis of CACs by XAS and XES**

Electronic structures in real fuel cells under operating conditions should be quite different from the above-mentioned electronic structures in UHV. In order to fully understand the ORR mechanism, CAC cathode in the membrane electrode assembly (MEA) configuration for PEMFC must be investigated under water-filled, pH-controlled and biased conditions, wherein photoelectrons might not escape from the system. Therefore we constructed a new *in situ* photon in / photon out spectroscopy (PIPOS) system with a varied line spacing monochromator at the University of Tokyo soft X-ray beamline at SPring-8. A photograph of the system is shown in Fig. 11(a). In addition, we designed an electrochemical half cell with a 150 nm thick SiC window, as shown in Fig. 11(b). So far, we have achieved a high-energy resolution of about 120 meV for emitted photons at 640 eV of incident photon energy by observing *d-d* excitation in Mn 2*p*-3*d* using a MnO sample [25]. The *in situ* XES analysis results on CAC PEMFC will be discussed elsewhere.

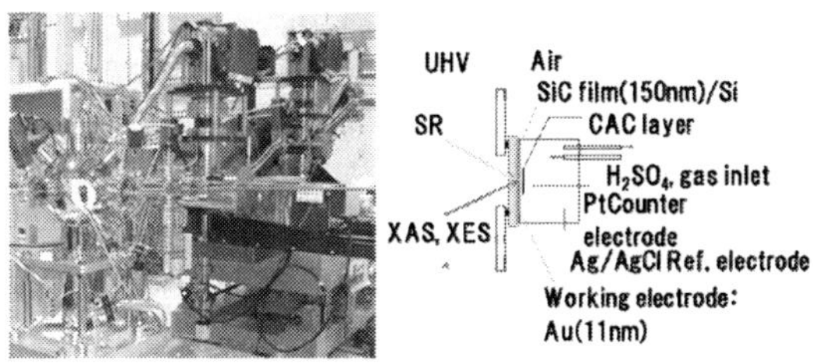

(a) (b)

Figure 11. *In situ* XES system for fuel cells during operation at BL07LSU of SPring-8.

CONCLUSIONS

We have investigated the electronic and chemical structures of non-Pt carbon alloy catalysts for polymer electrolyte membrane fuel cells using HXPES, XAS, and XAFS measurements to elucidate the ORR mechanism, in order to further improve ORR activities. It was found that acid washing reduces the cobalt concentration in the CACs but does not change either the carbon structure or the chemical state of nitrogen. Because the cobalt content in the CACs is only 0.03 at.% within the probing depth of the photoelectrons, we suggest that cobalt is not responsible for the ORR activity in the CACs. The results indicate that the carbon atoms adjacent to graphitic nitrogen at a zigzag edge of graphite may be the ORR active site, which is in good agreement with first-principles calculations, and that zigzag edges detected as an XAS shoulder component are well evolved for catalysts pyrolyzed at 600 °C showing the maximum ORR activity.

These results suggest a strategy for the further improvement of carbon-based ORR catalyst performance, which involves increasing ORR active sites by increasing the zigzag edge density itself along with the graphitic nitrogen concentration at the zigzag edge, and also by improving the electrical conductivity in the cathode. Therefore, robust graphite-like structures with adequate size to preserve the zigzag edge density and electrical conductivity are needed to improve both the ORR performance and the durability of PEMFC. Based on these results, we realized significantly improved ORR performance, up to about 70% of that observed by using Pt catalysts, by optimizing the pyrolysis and post-treatment conditions. In order to analyze the electronic structure of CACs during operation, we constructed a new *in situ* soft X-ray emission spectroscopy system having a high-energy resolution (ΔE) of 120 meV at 640 eV of photon energy at the University of Tokyo beamline BL07LSU in SPring-8.

ACKNOWLEDGMENTS

This work was carried out under the financial support from the New Energy and Industrial Technology Development Organization, Japan. HXPES experiments were performed in collaboration with K. Kobayashi, Y. Yamashita, H. Yoshikawa, and S. Ueda of NIMS and E. Ikenaga of JASRI. XAS experiments were performed with the support of K. Amemiya of KEK-PF.

REFERENCES

1. R. Jasinski, *Nature* **201**, 1212 (1964).
2. R. Bashyam and P. Zelenay, *Nature* **443**, 63, (2006).
3. J. Ozaki, N. Kimura, T. Anahara, and A. Oya, *Carbon* **45**, 1847 (2007). J. Ozaki *et al.*, *Electrochimica Acta* **55,** 1864 (2010).
4. R. Bashyam *et. al.*, *Nature* **443**, 63 (2006).
5. M. Lefèvre, E. Proietti, F. Jaouen, and J. Dodelet, *Science* **324**, 71 (2009).
6. Y. Shao, J. Sui, G. Yin, and Y. Gao, *Appl. Catal. B: Environ.* **79**, 89 (2008).
7. K. Gong, F. Du, Z. Xia, M. Durstock, and L. Dai, *Science* **323**, 760 (2009).
8. Y. Nabae, M. Malon, S.M. Lyth, S. Moriya, K. Matsubayashi, N.M. Islam, S. Kuroki, M. Kakimoto, J. Ozaki, and S. Miyata, *Carbon* **48**, 2613 (2010).
9. H. Dai *et al.*, *Chem. Phys. Lett.* **260**, 471 (1996).
10. M. Fujita, K. Wakabayashi, K. Nakada, and K. Kusakabe, *J. Phys. Soc. Jpn.* **65**, 1920 (1996).
11. K.S. Novoselov *et al.*, *Nature* **438**, 197 (2005).
12. T. Ikeda, M. Boero, S. Huang, K. Terakura, M. Oshima, and J. Ozaki, *J. Phys. Chem. C* **112**, 14706 (2008).
13. H. Niwa, M. Kobayashi, K. Horiba, Y. Harada, M. Oshima, K. Terakura, T. Ikeda ,Y. Koshigoe, J. Ozaki, S. Miyata, S. Ueda, Y. Yamashita, H. Yoshikawa, and K. Kobayashi, *J. Power Sources* **196**, 1006 (2011).
14. J. Casanovas *et al.*, *J. Am. Chem. Soc.*, **118**, 8071 (1996).
15. T. Toda *et al.*, *J. Electrochem. Soc.* **146**, 3750 (1999).
16. H. Niwa, K. Horiba, Y. Harada, M. Oshima, T. Ikeda, K. Terakura, J. Ozaki, and S. Miyata, *J. Power Sources* **187**, 93 (2009).
17. Sheng-Feng Huang, K. Terakura, T. Ozaki, T. Ikeda, M. Boero, M. Oshima, J. Ozaki, and S. Miyata, *Phys. Rev. B* **80**, 235410 (2010).
18. S. Entani *et al.*, *Appl. Phys. Lett.* **88**, 153126 (2006).
19. V.L. Joseph Joly *et al.*, *Phys. Rev. B* **81**, 245428 (2010).
20. M. Saito *et al.*, *the 215th Meeting of ElectroChemical Society*, Abstract #265, San Francisco.
21. Our recent theoretical analysis suggests that the edge carbons are indeed the source of the shoulder E* although the zigzag edge states may not be directly linked to it. The details will be published elsewhere.
22. L. Wu, Y. Nabae, S. Moriya, K. Matsubayashi, N. M. Islam, S. Kuroki, M. Kakimoto, J. Ozaki, and S. Miyata, *Chem. Commun.*, **46**, 6377-6379 (2010).
23. Y. Kobayashi, K. Fukui, T. Enoki, and K. Kusakabe, *Phys. Rev. B* **73**, 125415 (2006).
24. K. Sugawara, T. Sato, S. Souma, T. Takahashi and H. Suematsu, *Phys. Rev. B* **73**, 045124 (2006).

25. M. Kobayashi, Y. Harada, M. Oshima, Y. Senba, H. Ohashi, T. Tokushima, Y. Horikawa, and S. Shin, *VUVX 2010 Satellite Meeting* (2010), Saskatoon.

Mater. Res. Soc. Symp. Proc. Vol. 1318 © 2011 Materials Research Society
DOI: 10.1557/opl.2011.483

Quantum rods and dots-based structures & devices: Low cost aqueous synthesis and bandgap engineering for solar hydrogen and solar cells applications

Lionel Vayssieres
International Center for Materials NanoArchitectonics, National Institute for Materials Science, Namiki 1-1, Tsukuba, 305-0044 Japan

ABSTRACT

If one considers the largest and geographically balanced free natural resource available on Earth, that is seawater, and that more sunlight energy is striking our blue planet in one hour than all of our annual energy consumption, the direct solar-to-hydrogen conversion by photo-oxidation of seawater is a very straightforward and attractive solution for the production of hydrogen, as it is clean, sustainable and renewable. It offers an alternative solution to fossil-fuel-based energy sources and explains the tremendous interest in renewable, sustainable energy sources and materials for energy conversion. However, the materials requirements for water splitting and thus the direct solar-to-hydrogen generation are drastic. The materials must be stable in water, which rules out many classes of materials. They must also be stable under illumination against photocorrosion and their bandgap must be small enough to absorb visible light, but large enough not to "dissolve" once illuminated. Finally, their band edges must be positioned below and above the redox potential of hydrogen and oxygen, respectively. Bandgap energy and band-edge positions, as well as the overall band structure of semiconductors are of crucial importance in photoelectrochemical and photocatalytic applications. The energy position of the band edges can be controlled by the electronegativity of the dopants and solution pH, as well as by new concepts such as quantum confinement effects and the fabrication of novel hetero-nanostructures. Fulfilling those requirements while keeping the cost of the materials low is a tremendously difficult challenge, which explains why solar hydrogen generation is still in its infancy. Novel approach and latest development combining low cost aqueous synthesis techniques, vertically oriented metal oxide nanorods and quantum confinement effects probed by x-ray spectroscopies from synchrotron radiation is presented leading to stable and cost-effective visible-light-active semiconductors for seawater splitting, the holy grail of photocatalysis.

INTRODUCTION

Meeting the future demand of energy (which should double by 2050) and managing the environmental consequences of energy production is of major importance nowadays with the rise of emerging highly populated countries such as India and China. Considering that more energy of sunlight is striking earth in an hour than all of the energy consumed by humans in a year offers an alternative solution to fossil fuels-based energy source and explains the tremendous interest in renewable sustainable energy source and materials for energy conversion. Thus, finding new ways to power the future by making safer energy without creating extra CO_2 in the atmosphere to reduce urban air pollution and the building-up of greenhouse gases that threaten severe climate change (global warning) as well as simultaneously lowering dependence on foreign oil has become a crucial priority if not simply the most important problem facing humanity today. Indeed, the transition from fossil fuels to hydrogen is of revolutionary importance not only for its societal impact but also for the new discovery of materials for renewable energy [1]. New sciences and new technologies create new industries, which provide in turn job opportunities as well as better knowledge and living conditions [2]. The ages of ancient civilization are remembered by their materials. Truly, the present age will be known by the nanomaterials that will provide our society the necessary tools to supply sustainable energy source and energy source carrier while protecting its environment. Making the necessary revolutionary transition of energy from its fossil fuel beginning to its clean and renewable energy present and future is in our hands. However, the major problem to the widespread implementation of solar related energy has been clearly identified as the cost of generation of the renewable energy. The need of ultra-low cost purpose-built, functional materials with optimized geometry orientation and aspect ratio combined with inexpensive large scale manufacturing methods is the key to the success of solar related energy source. The design and large scale fabrication of ordered arrays consisting of advanced and well-defined building blocks such as quantum dots, nanorods, and nanowires is essential to the creation of new devices. However, manufacturing large area of such materials is a daunting material engineering challenge.

MATERIALS AND METHODS

Materials synthesis by chemical methods has emerged as one of the most versatile fabrication techniques for the in-depth study and rational delivery of high purity functional nanomaterials, engineered from the molecular to the microscopic scale at low cost and large scale. The latter attributes make this approach very attractive for industrial applications and represents the immediate answer to the enormous need for new materials research to develop technologies for renewable energy. Our approach is to fabricate innovative semiconductor nanostructures and nanodevices for efficient photoelectrochemical production of hydrogen under visible-light irradiation. The nanomaterials based on metal oxides, the most stable and versatile class of compounds in terms of physical and chemical properties, will be purposely designed at multi length scale (nano, meso, and micro) from low cost syntheses methods such as aqueous chemical growth [3-17] onto various substrates including the industrially required transparent conducting substrates and subsequently doped by chemical and/or sputtering/vapor deposition. Two- and three-dimensional nanostructures based on advanced and functional zero and one-dimensional building blocks such as quantum dots, quantum rods and nanowires, were subsequently optimized by selected processes consisting of chemical coating, specific heat treatment, and interfacial electrochemistry to fabricate efficient and stable nanodevices with optimum bandgap for visible-light absorption. Their optical, structural and electronic properties were thoroughly investigated at synchrotron radiation facilities. Simultaneously their photoelectrochemical and photocatalytic properties were probed at real scale with solar simulator bench set-ups to record their photoconversion efficiency to achieve the ideal bandgap profiling for efficient visible-light absorption. A constant and direct feedback between materials development and in-depth fundamental knowledge and experimental investigation allowed selecting, tuning and optimizing the properties of the materials.

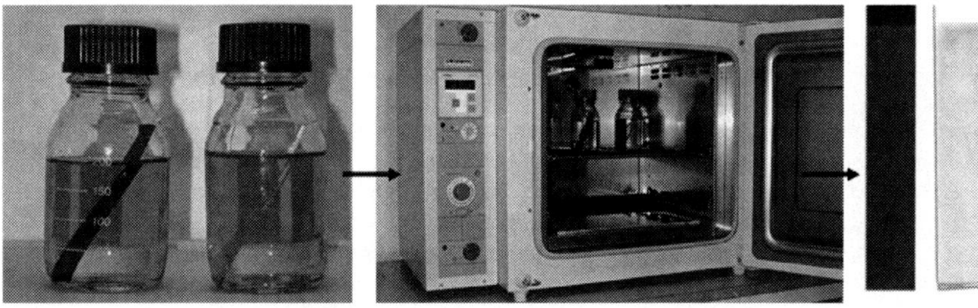

Figure 1. The aqueous chemical growth (ACG) thin film processing technique: immersion of substrates in aqueous solutions of metal salt precursors at room temperature (left), heat treatment in laboratory oven below 100 °C (center), and generation of nano-, meso- and micro-particulate arrays of large physical surface area (tens to hundreds of cm^2) onto various substrates (right).

The synthesis method we used originates from the findings of a thermodynamic model [18] based on the observations that divided and ultra-divided systems such as nanoparticle and/or colloidal dispersions are generally unstable with regard to the size and number of their constituents because the solid-solution interfacial tension, acting as a driving force, leads to a spontaneous reduction of the surface area to minimize the dispersion free enthalpy.
Such phenomenon known as surface energy minimization induces an increase in average particle size as a result of the reduction of the surface area at constant volume. For such a reason nanoparticle dispersions are considered thermodynamically unstable. Indeed, nanoparticles in suspension can undergo spontaneous (or timely) changes in size, shape, and/or crystal structure. These transformations occur because nanoparticles form under kinetic control during the initial steps of the nucleation and growth processes involved in the precipitation reactions. Further changes may occur during ripening to reach a thermodynamic equilibrium of minimized surface energy. The evolution in size, known as Ostwald ripening, involves dissolution-crystallization processes which lead to secondary growth processes and thus, the overall increase of the average particle size with time. This process originates from the difference of solubility between nanoparticles of different sizes – the smaller, the more soluble. Thus, the smallest will dissolve and re-crystallize on the larger ones.
As a result, the decrease of the total surface area of the system allows reducing the free enthalpy of formation of the solid. It is somehow possible to obtain dispersions consisting of very fine particles due to a very slow ripening, i.e. by growth limitation due to surfactants, and yet, the system is not in a stable state from a thermodynamic point of view, and consequently, the final particle size cannot generally be rationally controlled or predicted and the surface of the nanoparticles is being contaminated by the surfactant. Such effect becomes crucial while using nanoparticles for their surface properties such as in catalysis for instance where part of the actives sites may be inactivate due to the presence of remaining C, N, S from the surfactant. However, they can be thermodynamically stabilized if, by adsorption, the interfacial tension of the system becomes very low. This phenomenon, well known for (micro)emulsions, was for the first time quantitatively modeled and demonstrated for transition metal oxide nanoparticles. When the pH and the ionic strength of the aqueous precipitation/dispersion medium are

sufficiently high, the ripening of nanoparticles is avoided and their size can be monitored over an order of magnitude by tailoring solution pH and ionic strength.
A model based on Gibbs adsorption equation lead to an analytical expression of the water-oxide interfacial tension as a function of the pH and the ionic strength of the dispersion/precipitation medium. The stability condition, defined by a zero interfacial tension, corresponds to the chemical and electrostatic saturation of the water-oxide interface [19]. In such a condition, the density of charged surface groups reaches its maximum and further adsorption forces the surface area to expand and consequently, the size of nanoparticles to decrease. An empirical relation between the particle size and the conditions of precipitation was also established. This model accounts very well for the experimental results obtained from the aqueous precipitation of spinel iron oxide nanoparticles (magnetite, Fe_3O_4) in alkaline medium [20]. A general control of the metal oxide nanoparticle size when precipitated far from their point-of-zero-charge was expected and later confirmed on other metal oxide systems.
In addition, the nanocrystal structure could also be tailored as crystal phase transformations in aqueous solutions generally operate through a dissolution-recrystallisation process to fulfil surface energy minimisation. Indeed, when a solid offers several allotropic crystal phases and polymorphs, it is usually the one with the highest solubility and so the lowest stability (i.e., the crystallographic metastable phase), which precipitates first.
This is understood by considering the nucleation kinetics of the solids. At a given supersaturation ratio, the germ size is as small (and the nucleation rate as high) as the interfacial tension of the system is low. Thus, given that the solubility is inversely proportional to the interfacial tension, the precipitation of the most soluble phase and consequently the less stable thermodynamically, is therefore kinetically promoted. Due to its solubility and metastability, this particular phase is more sensitive to secondary growth and ageing, which leads to crystallographically more stable phases, essentially by heteronucleation.
It is typically the case of, for instance, aluminium hydroxide (*boehmite* to *bayerite* or *gibbsite*), titanium dioxide (*anatase* or *brookite* to *rutile*), or iron oxyhydroxides (*goethite*, *akaganeite*, or *lepidocrocite* to *hematite*). Ageing processes are delicate to control and the phase transformations appear within a few hours to a few days in solution, whence the resulting undesired mixing of allotropic phases and polymorphs.
However, by careful consideration of the precipitation conditions such phenomena may be avoided when nanosystems are precipitated at low interfacial tension. Thus, the stabilisation of metastable nanocrystal structures can be achieved at thermodynamically stable conditions.

Experimentally, the chemical fabrication method of advanced metal oxide nanostructures simply consists of heating homogeneous aqueous solutions of metal salts (e.g. $FeCl_3$, $Zn(NO_3)_2$...) in the presence of a substrate (e.g. glass, Si/SiO_2 wafers, sapphire, glass, coated glass, plastics...) at moderated temperatures (below 100°C) in a closed vessel (figure 1). Therefore, such technique does not require high-pressure containers and is also entirely recyclable, safe, and environmental-friendly since only water is used as solvent. Such a process avoids the safety hazards of organic solvents and their eventual evaporation and potential toxicity. In addition, since no organic solvents or surfactants are present, the purity of the materials is dramatically improved. Any residual salts (from the precursors decomposition and/or from the added ionic strength) are easily washed out by water due to their high solubility. In most cases, no additional heat or chemical treatments are necessary, which represent a significant improvement compared to surfactant, template or membrane based synthesis methods.

A full coverage of the substrate is obtained within few hours (and up to a couple of days in certain cases) provided that the heat capacity of water and surface coverage control are achieved by monitoring the synthesis time in the early stages of the thin film growth. Partial coverage is obtained within the first hours, which may be necessary for certain applications to adjust and tune the overall physical properties of devices (e.g. optical properties of multi-bandgap thin films). Such a technique is a multi-deposition technique and the growth of layer-by-layer of thin films is readily obtainable. The development of multilayer thin films of various morphologies and/or of various chemical compositions – i.e. composite, multi-bandgap, and doped thin films – is reached. The complete thin film architecture may thus be modeled, designed and monitored to match the requirements of the applications. In most cases, it does improve the physical and/or chemical properties of the devices. It also gives the capacity to create novel and/or improved thin film integrated devices.

Growing thin layers directly from the substrate does substantially improve the adherence and the mechanical stability of the thin film compared to the standard solution and colloidal deposition techniques such as spin and/or dip coating, chemical bathing, screen printing, or doctor blading. Moreover, given that such materials do precipitate in homogeneous solution from molecular scale compounds (i.e. condensed metal complex), they will grow on virtually any substrate (providing that the substrates are stable in aqueous solutions). It goes without saying that the overall mechanical stability of the thin films does vary from substrate to substrate, but in most cases, strong adhesion is observed. Scaling-up is easily feasible, and this concept and synthesis method is potentially applicable to all water-soluble metal ions likely to precipitate in aqueous solutions. Large-scale manufacturing at low cost is therefore achievable with such a technique. In addition to all these industrial related advantages, such a technique is also potentially very interesting due to the compatibility of water and aqueous solution to biological compounds.

Moreover, additional strategy to fully address the drastic requirements of novel functional nanomaterials development with higher performance and improved stability by atomic and molecular design and control is the ability to design metal oxide semiconductor based on vertically oriented anisotropic nanostructures (i.e. nanorods) with intermediate bands and highly quantized band structure such as quantum rods and dots to enable high efficiency in the visible range as well as tuning bandgap and band edges by quantum confinement effects. Such unique characteristics, combined with in-depth investigation and modeling of their electronic structure and large scale and low cost fabrication methods provide this research a substantial advance and (r)evolutionary prospect in the field of semiconductor technology and materials for solar energy conversion.

Such ideas is illustrated and demonstrated on design and fabrication of metal oxide quantum dots and quantum rods-based large semiconductor structures and devices as well as the in-depth characterization of their electronic structure performed at synchrotron radiation facilities [21-25] and applications for visible light-active semiconductors for direct solar hydrogen generation [26] and photovoltaics [27-31], as well as materials for sensors [32-35].

The attempt to improve the performance of existing materials by further development along the same lines (incremental increase), is reaching its limit. Indeed, very little possibility of further improvement and innovations is foreseen especially considering the environment when aiming at even higher efficiency. Given this situation, it is now imperative to develop novel functional materials which can produce real innovations with the aim of realizing a true renewable energy production. The necessity of materials development which is not limited to materials that can achieve their theoretical limits, but makes it possible to raise theoretical limits by changing the fundamental underlying physics and chemistry has become a critical issue. Materials R&D fully exploiting nanoscience has the greatest potential to reach such highly challenging goals. Indeed, recent reports of such advanced materials using quantum confinement effects have surfaced and triggered a renewed interest and new hopes for higher efficiency and visible light active materials for photovolatics and photocatalysis. One of the major limiting steps to the development of materials for solar photoconversion with high efficiency is the 1:1 ratio between the absorption of photons and the creation of electron-hole pairs in the materials. A process called impact ionization, known in the semiconductor field since the late 50s, allows the creation of more than one electron-hole pair per absorbed photons. Although, such a process is not efficient for bulk semiconductors (due to the high photon energy threshold), it becomes very attractive and very efficient when dealing with nanoscale semiconductors where the threshold photon energy to generate multiple electron-hole pairs could occur in the visible or near IR spectral region. Such a process is now called "multiple exciton generation".

Another quantum effect to efficiently absorb photons with energy lower than the bandgap and thus create electron-hole pairs to increase the efficiency in the visible part of the solar spectrum of photovoltaics and photocatalytic devices, is the so-called "intermediate band" materials. Figure 2 illustrates such a concept in quantum-dot based materials. The intermediate band arises from the confined electronic states of the electrons in the potential wells of the conduction band. Delocalization of the electrons within the intermediate band is achieved by increasing their density until the electron wave functions have significant overlap and thus become delocalized (i.e. Mott transition). Indeed, by creating an (intermediate) band between the valence band and the conduction band, interband carrier relaxation is substantially reduced (limited probability of multi-phonon interactions) and sub-bandgap photons can contribute to the overall photocurrent generation. It is important to understand that with this concept, the introduction of energy levels within the bandgap of the semiconductor is not supposed to create non-radiative recombination centers (which usually reduce the performance instead of improving it). The main fundamental explanation is that the wave function associated with the electrons in the intermediate band has to be delocalized to improve efficiency. This is contrary to conventional non-radiative recombination centers, which are tightly bound.

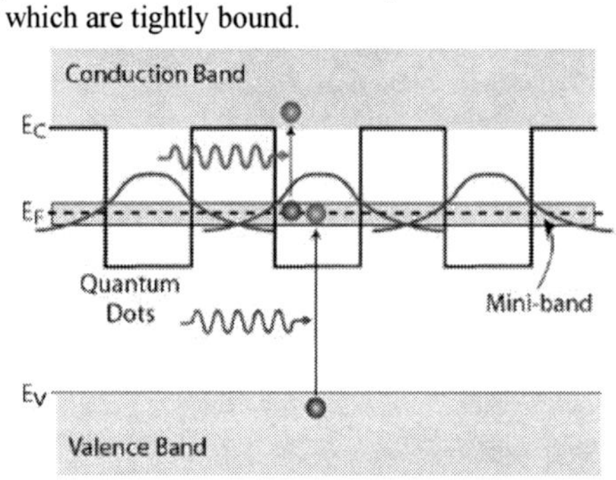

Figure 2. Schematic representation of the energetics and optical transitions of quantum dot-based structures. The intermediate band arises from the confined electronic states of the electrons in the potential wells of the conduction band which enables the absorption of photons of two different energies in addition to the bandgap absorption [36]

RESULTS AND DISCUSSION

The major innovations are the ability to develop at low cost and large scale novel metal oxide nanostructured materials with highly quantized band structure such as quantum rods and quantum dots to generate high energy conversion and full visible absorption as well as bandgap increase for enhanced stability against photocorrosion due to low-dimensional quantum confinement. The materials development include doped and undoped metal oxide vertically oriented nanorod arrays fabricated by low temperature chemical techniques and quantum dots sensitized quantum rods metal oxide vertically oriented arrays fabricated by low temperature aqueous chemical growth onto various transparent conducting oxide substrates.

Such unique characteristics, combined with in-depth investigation and modeling of their electronic and large scale and low cost synthesis methods provide a substantial advance and revolutionary prospect in the field of semiconductor technology and materials for energy conversion. Such an advanced knowledge of the electronic structure and structural properties has direct implication for fundamental understanding of the structure-property relationships as well as efficiency optimization of new nanodevices. The research focuses on simple transition metal oxides with well-established properties purposely-designed as ordered doped nanorods and/or functionalized by quantum dots using cost-effective methods to fabricate the ideal visible-light active semiconductors for efficient solar hydrogen and other nanodevices.

In-depth investigations of the electronic structure of such novel nanostructures have been carried out at synchrotron radiation facilities. Indeed, the new generation synchrotron radiation sources producing intensive tunable monochromatized X-ray beams have opened up new possibilities for soft-X-ray emission spectroscopy (SXES). The possibility to select the energy of the excitation has created an extra degree of freedom compared to the traditional spectroscopy pursued with high-energy electron or characteristic X-ray excitation. The energy selectivity makes it possible to perform resonant excitation, i.e., exciting to particular empty states. The introduction of selectively excited soft-X-ray emission has opened a new field of study by disclosing many new possibilities of resonant inelastic soft-X-ray scattering (RIXS).

Among the new tools available with this technique one can mention for instance site selectivity, low energy excitations in correlated materials, and fs dynamics and chemical bonding mechanism by detuning from resonance. Both synchrotron radiation based soft-X-ray absorption spectroscopy (XAS) which map the unoccupied states of the materials (conduction band) and soft-X-ray emission spectroscopy (XES) which map the occupied state (valence band) on a variety of nanostructured systems have yielded important information on the bandgap, Fermi energy, work function as well as other band edges characteristics. In addition, angle-resolved XAS and XES spectroscopy along with DFT calculation have also been carried out (figure 3-4) on the materials (thin films, powders and even aqueous suspensions (with the newly built cells) to probe the orbital character and symmetry of the conduction band (and valence band) of the semiconductor hetero-nanostructures [21-25]. With high resolution synchrotron radiation excitation, RIXS has emerged as a new source of information about electronic structure and excitation dynamics of nanomaterials. The selectivity of the excitation, in terms of energy and polarization, has also facilitated studies of emission anisotropy as well as for probing the optical transitions in transition metal oxides, especially *dd* transitions which are (in other spectroscopies such EELS or optical spectrophotometry) dipole forbidden and thus of very low intensity (figure 4).

Synchrotron based spectroscopic investigations of 1-D nanomaterials consisting of designed bundled quantum rods of hematite reveal a significant increase of the bandgap compared to bulk (polycrystalline and single crystal) samples (figure 3-4). Such finding is successfully attributed to a 1-D (lateral dimension) quantum confinement effect. Such a conclusion strongly suggests that such designed nanomaterials would meet the bandgap and band edges criteria (figure 4) for the photocatalytic oxidation of water without an applied bias. Indeed, an increase in bandgap of 0.3 eV found in the quantum rod arrays would bring the top of the conduction band edge of hematite at an energy level very similar to the level of H^+/H_2. The outcome of such a result is of great importance for the direct solar production of hydrogen as it may thus possible to evolve oxygen (and hydrogen) with a very small applied bias from iron oxide photoanodes [26].

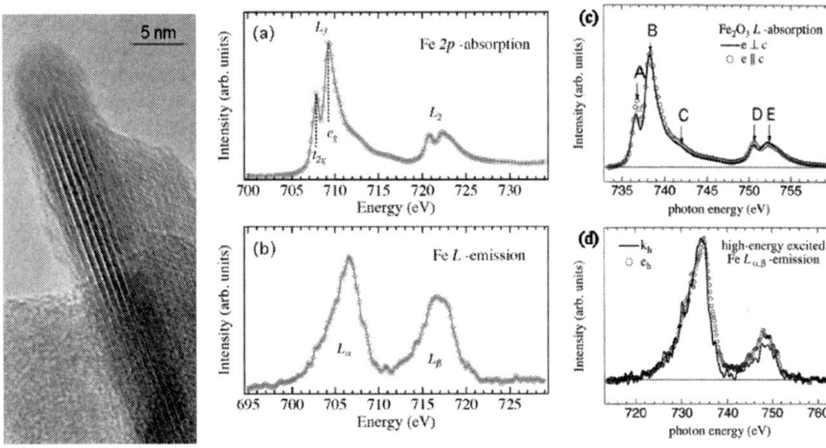

Figure 3. HRTEM image of an hematite quantum rod and Fe L-edge absorption and normal L-emission spectra of hematite quantum rod arrays (a, b) and single crystal (c, d) [37]. L-emission spectrum of the quantum rod arrays shows a substantially higher L_β/L_α branching ratio compare to the single crystal suggesting an upward shift of the conduction due to quantum confinement effect in the lateral dimension.

Quantum dot solar cells as well as inorganic and metal sensitization of semiconductors are efficient ways to tailor optical absorption and photoelectrochemical efficiency of semiconductors. To develop truly visible-light active materials, the fabrication of hetero-nanostructures of iron oxide was carried out. These novel structures consist of metal oxide quantum dots grown onto quantum rods of hematite. The development of such advanced materials is still in the preliminary stage but shows very promising results in terms of visible-light absorption. Figure 5 shows high resolution field emission scanning electron images along with the UV-Vis optical properties of quantum dot/quantum rod iron oxide hetero-nanostructure arrays. A very strong absorbance is found in the entire visible range and the absorption profile is an excellent match to the solar spectrum. The electronic structure investigation by angle resolved XAS and XES shows excellent bandgap engineering for water splitting applications.

Additional studies include the effect of size, not only on the electronic structure (i.e. quantum confinement) but also on the chemical properties (i.e. surface chemistry) of metal oxide nanostructures. For instance, we have shown that when the size of spinel iron oxide nanoparticles decreases from 12 to 7.5nm, the point of zero charge (PZC) changes from 6.3 to 7.0 (i.e. acidic to neutral surface). An additional size decrease to 3.5 nm further increased the PZC to 8.4, that is a basic surface [38]. Such results are of great importance for photovoltaics and photocatalysis as the surface chemistry of metal oxide affect its open circuit potential and chemical reactivity.

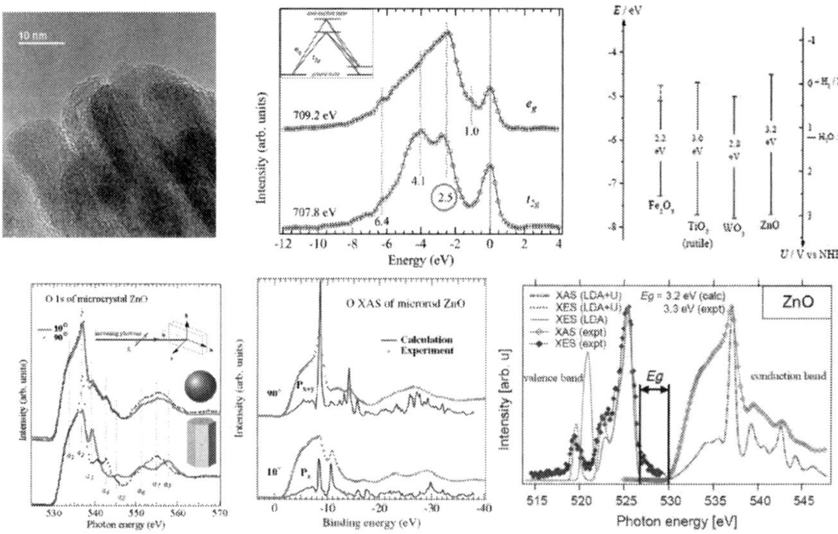

Figure 4. (top left) High-resolution electron image of a bundle of α-Fe_2O_3 quantum rods synthesized from aqueous solutions; **(top center)** RIXS spectra showing a 2.5 eV bandgap due to 1-D confinement for the quantum rods [21]; and **(top right)** schematic representation from the literature values of the conduction and valence band levels of large band gap semiconductors compared to the redox potentials of oxygen and hydrogen evolution. The increase of the bandgap of iron oxide quantum rods is shown. (**bottom left**) O K-edge polarization dependent XAS of ZnO isotropic and oriented anisotropic morphology [25]. The inset illustrates the experimental XAS geometry, where *a, b* axes defined the sample-surface plane, *c* axis was along the growth direction of the ZnO rods, ***E*** is the polarization of the incoming photons, and θ represents the incident angle to the sample surface: 10° ($\boldsymbol{E} \parallel c$, lines) and 90° ($\boldsymbol{E} \perp c$, dots); **(bottom center)** calculated polarization dependent O 1s XAS of wurzite bulk ZnO, where energy is referred to valence-band maximum; experimental (dots) and calculated (lines) O K-edge XAS spectra of ZnO as a function of the polarization angle. The calculated XAS spectra feature the partial density of states and the transition probability to O $2p_{x+y}$ and $2p_z$ states with a Lorentzian broadening equivalent to the experimental spectral resolution (0.2 eV); (**bottom right**) Experimental O 1s XAS and XES spectra of ZnO nanostructures and LDA+U calculation of bulk ZnO describes d-states localization, and thereby also the Znd-O-p hybridization. Eg is the electronic bandgap obtained by such spectroscopic measurements [23].

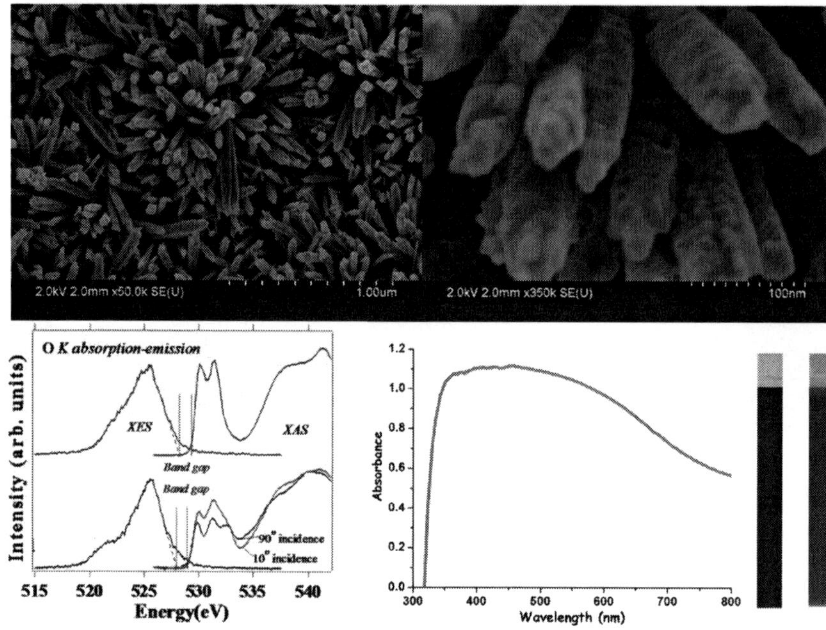

Figure 5. Field-emission SEM images of vertically oriented arrays of iron oxide quantum-rod sensitized quantum-dot hetero-nanostructures (**top**); O-K absorption and emission spectra of iron oxide-based hetero-nanostructures (**bottom left**); UV-Vis absorption spectrum of iron oxide based hetero-nanostructures showing full visible light absorption profile (**bottom center**); Still image of doped (black) and undoped (red) iron oxide photoanodes (**bottom right**).

CONCLUSIONS

Materials nanotechnologies based on chemical fabrication approaches is one of the immediate answer to the enormous need for cost-effective new materials for energy, electronics, and health. The ability to produce at low cost large three-dimensional patterned arrays of transition metal oxides consisting of anisotropic nanoparticles with controlled orientations onto various substrates without the use of templates, surfactants, or applied external fields was demonstrated. This approach to nanostructured arrays contributes not only to the cost-effective large scale development of smart and functional nanostructures and devices, but also to a better fundamental understanding of the physical and chemical properties of isotropic and anisotropic nanostructures such as quantum dots and rods. The major outcome is the capability of fabricating and processing quantum confined materials at low cost and large scale with bandgap and band edges relevant to induce water photo-oxidation at neutral pH and without sacrificial agent.

Indeed, a new approach using smart materials design and quantum confinement effects along with a simple and cost effective fabrication of advanced metal oxide materials for direct solar-to-hydrogen generation by photocatalytic oxidation of water under visible-light conditions was described in this chapter. This new generation of visible-light active iron oxide-based semiconducting nanostructures represents excellent candidates for mass production of efficient photocatalysts. They comply with the drastic requirements of photocatalysts for water oxidation in terms of stability (in water and under illumination), in terms of light absorption (visible-light active), in terms of safety (non-toxic) and in terms of cost (low fabrication and operating costs). More over, it demonstrates once again the advances of the application of nanoscience and nanotechnology in enabling the future hydrogen economy by fabricating novel visible light active materials by novel concepts and simultaneously reducing their cost of fabrication. The use of such new materials may render solar hydrogen generation commercially viable (e.g. the US department of Energy has set a price goal of 2-3US$ per kg of hydrogen which include production, delivery, and dispensing for hydrogen to compete with gasoline for passenger vehicles). It goes without saying that additional academic and industrial R&D is required for the complete optimization of those nanostructures in terms of overall efficiency, doping density, rod aspect ratio and density, and long term stability as well as their large scale fabrication and indoor/outdoor testing as $10x10cm^2$ modules and panels. Nevertheless, they will certainly pave the way to a new generation of low cost materials and devices from which one can easily imagine the creation of off-shore power plant dedicated to renewable direct solar-to-hydrogen fuel generation (using existing facilities for gas and oil) where seawater would be use as the sole electrolyte feedstock and hydrogen and oxygen generated and transported to the main land via gas pipelines. After all, the most abundant and freely available natural resources of this blue planet are the sunlight and seawater. Current and future studies include novel materials development as well as the effect of size on the surface chemistry of metal oxides (and thus on their catalytic properties) which has been elucidated on Maghemite (i.e. showing the reversal of the surface acidity from acidic to neutral to basic was demonstrated by size effect. Finally, the quantum size effect on the orbital character of metal oxide nanoparticles is also being investigated to improve the dynamics, transport and electrical properties of photogenerated carriers. Preliminary results show excellent trend for TiO_2 nanostructures.

ACKNOWLEDGEMENTS

This work was performed at the World Premier International (WPI) Center for Materials Nanoarchitectonics (MANA), National Institute for Materials Science (NIMS), Tsukuba, Japan which is supported by the Japanese Ministry of Science and Education (MEXT).

REFERENCES

1. *On Solar Hydrogen & Nanotechnology,* L.Vayssieres editor (John Wiley & Sons, 2009)
2. F. Rosei *et al.*, *Adv. Mater.* 20(24), 4627-4640 (2008)
3. L.Vayssieres, *Appl. Phys. A* 89(1), 1-8 (2007)
4. L.Vayssieres, *Int. J. Nanotechnol.* 4, 750-775 (2007)
5. L.Vayssieres, *C.R. Chimie* 9(5-6), 691-701 (2006)

6. L.Vayssieres and M. Graetzel, *Angew. Chem. Int. Ed.* 43(28), 3666-3670 (2004)
7. L.Vayssieres, *Int. J. Nanotechnol.* 1(1-2), 1-40 (2004)
8. L.Vayssieres, *J. Phys. Chem. B* 107(12), 2623-2625 (2003)
9. L.Vayssieres, *J. Nanosci. Nanotech.* 1(4), 385-388 (2001)
10. L.Vayssieres *et al.*, *Pure Appl. Chem.* 72(1-2), 47-52 (2000)
11. L.Vayssieres *et al.*, *Chem. Mater.* 13(2), 233-235 (2001)
12. L.Vayssieres *et al.*, *Nano Lett.* 2(12), 1393-1395 (2002)
13. L.Vayssieres, *Adv. Mater.* 15(3), 464-466 (2003)
14. L.Vayssieres *et al.*, *Chem. Mater.* 13(12), 4395-4398 (2001)
15. L.Vayssieres *et al.*, *J. Phys. Chem. B* 105(17), 3350-3352 (2001)
16. Y. Yang *et al.*, *Physica E* 40(4), 859-865 (2008)
17. L. Rabenberg and L.Vayssieres, *Microsc. Microanal.* 9(2), 402-403 (2003)
18. L.Vayssieres, Précipitation en milieu aqueux de nanoparticules d'oxydes: Modélisation de l'interface et Contrôle de la croissance, Thèse de doctorat CHIMIE 1995PA066747, Université Pierre et Marie Curie, Paris, France (1995), pp. 1-145
19. L.Vayssieres, *Int. J. Nanotechnol.* 2(4), 411-439 (2005)
20. L.Vayssieres *et al.*, *J. Colloid Interface Sci.* 205(2), 205-212 (1998)
21. L.Vayssieres *et al.*, *Adv. Mater.* 17(19), 2320-2323 (2005)
22. C. Persson *et al.*, *Microelectron. J.* 37(8), 686-689 (2006)
23. C.L. Dong *et al.*, *Phys. Rev. B* 70, 195325 (2004)
24. A. Henningsson *et al.*, *Adv. Quant. Chem.* 47, 23-36 (2004)
25. J.-H. Guo *et al.*, *J. Phys.: Condens. Matter* 14(28), 6969-6974 (2002)
26. L.Vayssieres,"Quantum confined visible-light active nanostructures for direct solar hydrogen generation" in *On Solar Hydrogen & Nanotechnology,* L.Vayssieres ed. (John Wiley & Sons, 2009), chapter 17, pp. 523-558
27. N. Beermann *et al.*, *J. Electrochem. Soc.* 147(7), 2456-2461 (2000)
28. T. Lindgren *et al.*, *Sol. Energy Mater. Sol. Cells* 71(2), 231-243 (2002)
29. T. Lindgren et al., "Photo-oxidation of water at Hematite electrodes", in *Chemical Physics of Nanostructured Semiconductors,* A. I. Kokorin and D. W. Bahnemann editors (VSP, Utrecht, The Netherlands, 2003), Chapter 3, pp. 83-110
30. K. Keis *et al.*, *Nanostruct. Mater.* 12(1-4), 487-490 (1999)
31. K. Keis *et al.*, *J. Electrochem. Soc.* 148(2), A149-155 (2001)
32. L.Vayssieres and X.W. Sun, *Sensor Lett.* 6(6), 787-791 (2008)
33. J.X. Wang *et al.*, *Nanotechnology* 17(19), 4995-4998 (2006)
34. L.Vayssieres, *Chem. Sensors* 20(B), 324-325 (2004)
35. L.Vayssieres, "3-D bio-inorganic arrays", in *Chemical Sensors VI: Chemical and Biological Sensors and Analytical Methods*, C. Brukner-lea, P. Vanysek, G. Hunter, M. Egashira, N. Miura, and F. Mizutani editors (*The Electrochemical Society*, Pennington, NJ, 2004), pp. 322-343
36. *Basic research needs for solar energy utilization,* Report of the Basic Energy Sciences Workshop on Solar Energy Utilization, April 18-21, 2005
37. L.-C. Duda *et al.*, *J. Electron Spectrosc. Relat. Phenom. 110-111(1-3)*, 275-285 (2000)
38. L.Vayssieres, *J. Phys. Chem. C* 113(12), 4733-4736 (2009)

Real-Time Studies of Evolving Thin Films and Interfaces

Mater. Res. Soc. Symp. Proc. Vol. 1318 © 2011 Materials Research Society
DOI: 10.1557/opl.2011.484

Formation of Irregular Al Islands by Room-Temperature Deposition on NiAl(110)

Dapeng Jing[1,6], Yong Han[5], Bariş Ünal[2,6,*], J. W. Evans[3,4,6] and P. A. Thiel[1,2,6]
Departments of [1]Chemistry, [2]Materials Science & Engineering, [3]Mathematics, and [4]Physics & Astronomy, Iowa State University, Ames, Iowa 50011
[5]Institute of Physical Research and Technology, Iowa State University, Ames, Iowa 50011
[6]Ames Laboratory - USDOE, Iowa State University, Ames, Iowa 50011

ABSTRACT

STM studies reveal that irregular non-equilibrium two-dimensional Al islands form during deposition of Al on NiAl(110) at 300 K. These structures reflect the multiple adsorption sites and diffusion paths available for Al adatoms on the binary alloy surface, as well as the details of inhibited edge diffusion and detachment-attachment kinetics of Al adatoms for numerous distinct step edge configurations. We attempt to capture these features by multi-site lattice-gas modeling incorporating DFT energetics for adatoms both at adsorption sites and transition states. This formulation enables description and elucidation of the observed island growth shapes.

INTRODUCTION

A rich variety of complex far-from-equilibrium morphologies are known to form in homoepitaxial growth of single-component metal films (A on A) at lower deposition temperatures, T. This behavior occurs despite the feature that these systems have very simple equilibrium states [1,2]. The complex morphologies result from kinetic limitations on step edge diffusion or interlayer diffusion on the time-scale of film growth. Similar behavior is observed in heteroepitaxy (A on B) at least for lower lattice misfit, but distinct strain-induced features generally occur for higher misfit [3]. Rather than the single-component substrates employed in the above studies, our focus here is in the use of binary alloy substrates where a variety of far-from-equilibrium structures should also form under low-T deposition. We are motivated by the potential for these systems to provide new opportunities to guide the creation of thin film nanostructures with desired properties. Interestingly, despite this potential, well-controlled ultra-high-vacuum (UHV) analyses of film growth on binary alloys (A or B on BC) are rare [4-7].

In this paper, we consider submonolayer deposition of Al on NiAl(110) at low $T = 300$ K, corresponding to far-from-equilibrium growth with effectively frozen substrate dynamics. However, it is appropriate to comment on a previous study of Al deposition on NiAl(110) at high $T \approx 900$ K, where the NiAl substrate is actually quite dynamic or "reactive" [4]. In the initial stages of Al deposition at around 900 K, half the deposited Al replaces excess Ni residing at Al sites in the bulk. This displaced Ni diffuses to the surface and combines with the other half of the depositing Al to form a new stoichiometric NiAl(110) alloy adlayer [4]. Once the excess Ni concentration in the substrate is sufficiently reduced, the system turns to another mechanism involving bulk defects to bring Ni to the surface to perpetuate alloy growth [4]. One could imagine that at lower T, the depositing Al could at least extract excess Ni at Al sites in the surface layer. However, at 300 K, our density functional theory (DFT) analysis indicates that even this process is inactive, i.e., Al is effectively deposited on an inert substrate.

Our development of atomistic-level modeling for Al deposition on NiAl(110) is aided and validated by comparison with experimental STM observations. There are complications for

Al/NiAl(110) not seen, e.g., in Al homoepitaxy, such as multiple types of adsorption sites and diffusion pathways. Thus, a realistic description requires multi-site lattice-gas (LG) modeling [5,6]. Furthermore, an accurate treatment of far-from-equilibrium growth kinetics requires precise determination of diffusion barriers including those for edge diffusion and attachment-detachment for many local step edge configurations. This is achieved here utilizing high-level DFT to probe energetics for adatoms not just at adsorption sites but also at transition states for diffusion [5,6]. Model behavior can then be assessed via kinetic Monte Carlo (KMC) simulation.

EXPERIMENTAL DETAILS AND RESULTS

The experiments were performed in a UHV chamber with a base pressure of $\leq 4 \times 10^{-11}$ Torr. The chamber is equipped with an Omicron VT-STM as well as other standard facilities for sample preparation and characterization (Sputter gun, LEED, AES and XPS). The NiAl(110) single crystal was grown using the Bridgman technique. A sample was polished using standard metallographic techniques and mounted on an Omicron heater. The sample was cleaned by repeated cycles of 20 min of Ar^+ sputtering (1.5 keV at 300 K) followed by annealing to 1250 K for 1 h. This resulted in a clean surface with broad terraces with an average width of 200 nm. Al was evaporated from a homemade deposition source. The base pressure was below 9×10^{-10} Torr during deposition and the purity of the film was checked by XPS. Flux calibration of the source was achieved by measuring the coverage directly from the STM images in the submonolayer regime. Sample orientation was checked by co-deposition of Ag which forms bilayer islands elongated in the [001] direction [7]. STM data was analyzed using WSxM software.

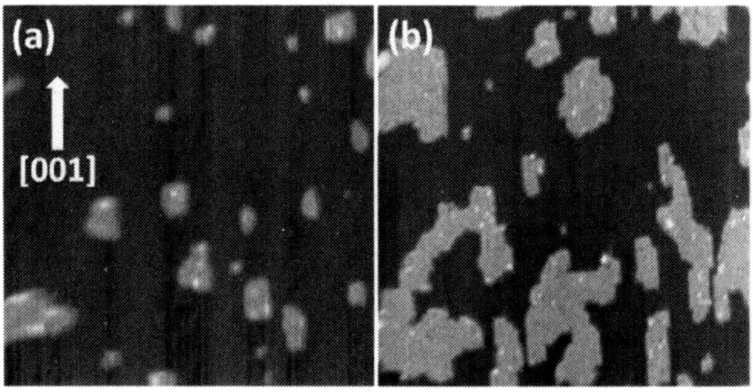

Figure 1. STM images (50×50 nm^2) of Al deposition on NiAl(110) at 300 K: (a) 0.1 ML with F = 0.0033 ML/s; (b) 0.36 ML with F = 0.0030 ML/s.

Al deposition at 300 K with flux F = 0.003 ML/s produces islands with a density of $\sim1\times10^{-2}$ nm^{-2} at ~0.1 ML. However, this value mainly reflects the surface defect density as island formation is dominated by heterogeneous nucleation at such surface defects rather than by homogeneous nucleation [5]. The prominence of heterogeneous nucleation seems to be a general feature for deposition on alloy surfaces, as opposed to deposition on single-component single-crystal surfaces [7]. Defects, such as those due to surface enrichment of one component, should generally be expected for alloys. In contrast to island density, island growth shapes are likely not significantly affected by the details of nucleation (e.g., the heterogeneous versus homogeneous pathway). From the STM images shown in figure 1, we see that smaller Al islands tend to be fairly compact and geometric with a propensity for [001] steps, but islands become irregular during growth. From these observations, it is clear that edge diffusion of aggregated Al adatoms is inhibited at 300 K. Provided that edge diffusion is somewhat operative, one expects a transition from compact shapes for smaller islands (which, in general, are growth rather than equilibrium shapes) to irregular shapes for larger islands. The latter reflects the longer distance which edge atoms must travel to reach more highly-coordinated kink sites [2,8].

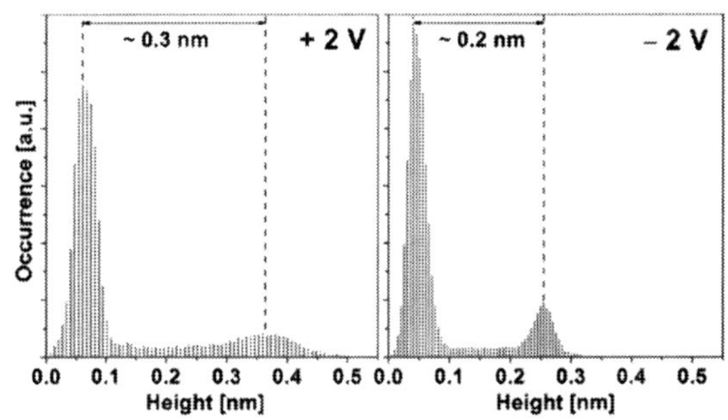

Figure 2. Tip-bias dependence of island height as reveals through film height distributions at 0.26 ML: lower (upper) peaks comes from the substrate (islands).

Another feature revealed in these studies is a strong tip bias-dependence of the measured apparent island height. This measured height is ~0.2 nm for a –2 V tip bias versus ~0.3 nm for a +2 V tip bias. Consistent results for this behavior come from either line scans or film height distributions. Data for the latter are shown in figure 2.

MULTI-SITE LATTICE-GAS (LG) MODEL DEVELOPMENT

DFT analysis, as well as comparison with STM observations, is invaluable in guiding model development. Specifically, DFT provides key energetic input parameters. Our analysis uses the plane-wave based VASP package [9] within the generalized gradient approximation using the PBE functional [10], and applying the PAW approach to treat electron-ion interactions [11].

To develop a detailed atomistic-level understanding of Al/NiAl(110), we first consider the binding and diffusion of an *isolated* Al atom on NiAl(110), the behavior of which is illustrated schematically in figure 3(a). Al prefers the short bridge site between two Ni aligned in the [100] direction (which is vertical in our schematics). This site is denoted as Ni-br below. Isolated Al diffuses preferentially in the [001] direction between Ni-br sites over an on-top Ni site transition state (TS) with a barrier of $E_d = 0.30$ eV. Al can also hop in the [110] direction from the on-top Ni site (Ni-top) over an Al short bridge (Al-br) site with a barrier of $E_d = 0.51$ eV. In contrast to behavior for isolated adatoms, diffusion of *aggregated* Al along island edges occurs predominantly via diagonal hops where the TS is a bridge-site between diagonally neighboring Ni and Al atoms (NiAl-br). We will describe the associated barrier determination below.

We consider a multi-site LG model where Al atoms within islands can populate both the Ni-br and Al-br sites, but we do not allow population of other sites. (See below for a critique of this limitation.) To prescribe the associated *adlayer thermodynamics*, we assess from DFT the interactions between Al adatom pairs at both types of sites for various separations. See figure 3(b). First, we describe consequences for island structure in this model incorporating just these pair interactions. The energy per adatom is lower for complete "dense" adlayers of Al populating *both* Ni-br and Al-br sites than for complete "dilute" adlayers populating just Ni-br or just Al-br sites: −4.68 eV vs. −4.00 eV (Ni-br) or −3.36 eV (Al-br) per Al adatom. This behavior tracks DFT energetics for complete adlayers for these three possibilities. The preference for "dense" over "dilute" islands reflects the presence of strong attractions between Al at diagonally neighboring Ni-br and Al-br sites which off-set the cost of populating the less favorable Al-br sites. Finally, we note that knowledge of these pair interactions also allows prediction of other key thermodynamic features. For example, for dense Al islands, one finds that [100]-oriented steps with Al on Ni-br sites are slightly favored over diagonal or horizontal steps.

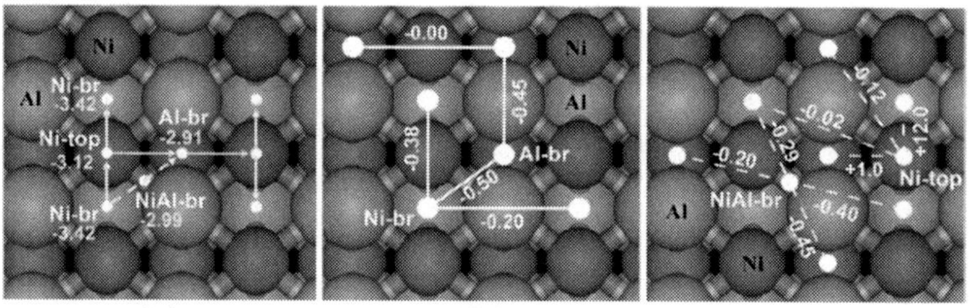

(a) (b) (c)

Figure 3. (a) Al adsorption energies in eV at various sites (binding < 0) and Al diffusion paths; pair interactions in eV (attractions < 0) between Al adatoms: (b) with both Al at adsorption sites; (c) with one Al at a TS and the other at an adsorption site.

Correct treatment of non-equilibrium *growth kinetics* also requires an accurate description of diffusion not just for isolated adatoms, but also for edge diffusion and attachment-detachment processes for all relevant pathways and for all local island edge configurations. Our model includes hops between Ni-br and Al-br sites with barriers determined from $E_{act} = E_{TS} - E_{init}$, where E_{init} (E_{TS}) is the total energy in the initial (transition) state. Both energies are obtained from a sum of the relevant adsorption energy and all pair interaction energies. The adsorption energies are provided in figure 3(a). Pair interactions relevant for the initial state are provided in figure 3(b). However, to determine E_{TS}, we must also assess a distinct class of pair interactions with one adatom at a TS and another at nearby adsorption sites. See figure 3(c). One caveat with the above formulation is that the true TS could be an initial or final unstable Al-br site (rather than a Ni-top or NiAl-br site), a feature checked for each Al hop. Finally, Arrhenius hop rates are selected with the above barriers and with prefactors $\nu = 10^{13}$/s. Detailed-balance is automatically satisfied. These rates provide input to KMC simulation of our atomistic multi-site LG model in which adatoms can reside at both Ni-br and Al-br sites, and where deposition and hopping are implemented with the appropriate relative probabilities. This formulation provides a powerful general strategy for modeling kinetics in epitaxial alloy thin film systems.

Finally, we briefly comment on some issues and limitations for our modeling. DFT, which is used to guide selection of energetic input parameters, has its limitations, and adjusting some interactions can better describe finer details of island shapes. We do not include trio, quartet, etc. adatom interactions, noting that the pairwise contributions capture key adlayer thermodynamics. Also, exchange with the substrate and concerted terrace diffusion processes are not included, but we have checked that they are energetically prohibitive at 300 K. More significantly, we neglect strain effects although dense Al islands are subject to compressive strain due to lattice mismatch with NiAl(110). Local strain relief induces Al at Al-br sites to sit higher than those at Ni-br. We also neglect alternative energetically-competitive monolayer island structures with so-called "square-hex motifs", where 3-fold-type sites in addition to Ni-br and Al-br sites are populated. These structures have been discussed for Au and Ag on NiAl(110) [6,7].

MULTI-SITE LG MODELING: KMC SIMULATION RESULTS

Results from KMC simulations of the model described above are presented here and compared with experiment. Our central focus is description of the nucleation and growth of Al islands. This process is mediated by strongly anisotropic terrace diffusion. The most stable dimer is vertically aligned on Ni-br sites with a binding energy of $E_b = 0.38$ eV (the vertical dimer on Al-br sites being destabilized by a large adsorption site penalty). This large E_b would imply irreversible island formation for homogeneous nucleation [2]. However, as noted above, heterogeneous nucleation dominates in this system, the enhanced stabilization of larger islands implying effectively irreversible island formation at 300 K. The simulated island density at 300 K from our model including only homogeneous nucleation (i.e., no defects) is far below experiment. This result indicates the dominance of heterogeneous nucleation in experiment.

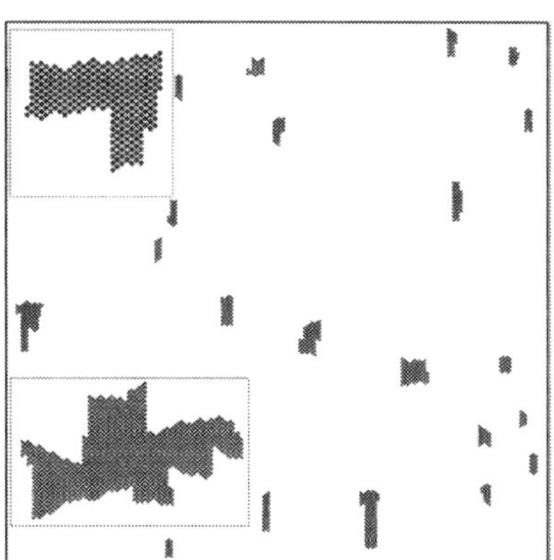

Figure 4. Simulation (size: 43×43 nm^2) of growth of multiple Al islands at 300 K with $F = 0.0033$ ML/s. Insets show structures of two larger islands.

Our simulations do recover trends in experimental island shapes despite the model limitations described above. Note that the island density determines the typical area of "capture zones" surrounding islands. This area controls the rate at which islands incorporate aggregating atoms, which in turn has some influence on island shape (higher aggregation rates inhibiting shape relaxation). Thus, to appropriately describe experimental conditions, we seed the system with a suitable density of Al island nuclei. Then, islands grow irreversibly about these during subsequent deposition. Results shown in figure 4 are consistent with experiment, i.e., a tendency for [001] steps at least in smaller islands, and the development of irregular growth of larger islands. Note that in simulations incorporating only the vertical and horizontal diffusion pathway for isolated Al, edge diffusion becomes effectively inoperative producing highly ramified fractal islands. Edge diffusion is activated via diagonal hops, especially along vertical edges ($E_{edge} \sim 0.5$ eV), facilitating formation of the island structures seen in experiment. Thus, multi-site multi-diffusion-path LG modeling is essential to capture growth behavior in this system.

SUMMARY

Deposition of Al at 300 K on NiAl(110) produces far-from-equilibrium dense monolayer islands. Multi-site LG modeling incorporating DFT-guided energetics for both interactions and barriers allows a realistic treatment of far-from-equilibrium growth in this complex system. This approach has general applicability for single- and multi-component epitaxial systems.

ACKNOWLEDGMENTS

This work was supported by the Division of Materials Sciences, US Department of Energy (USDOE) – Basic Energy Sciences. Computational resources were provided by the USDOE at NERSC. The work was performed at Ames Laboratory which is operated for the USDOE by Iowa State University under Contract No. DE-AC02-07CH11358.

*Current address: Department of Chemical Engineering, MIT, Cambridge, MA 02139.

REFERENCES

1. T. Michely and J. Krug, *Islands, Mounds, and Atoms* (Springer, Berlin, 2004).
2. J.W. Evans, P.A. Thiel, and M.C. Bartelt, Surf. Sci. Rep. **61**, 1 (2006).
3. R.Q. Hwang and M.C. Bartelt, Chem. Rev. **97**, 1063 (1997).
4. J.P. Pierce, N.C. Bartelt, and K.F. McCarty, Phys. Rev. Lett. **99**, 026101 (2007).
5. Y. Han, D. Jing, B. Unal, P.A. Thiel, and J.W. Evans, Phys. Rev. B., submitted (2011).
6. T. Duguet, Y. Han, C. Yuen, D. Jing, B. Unal, J.W. Evans, and P.A. Thiel, Proc. Nat. Acad. Sci. **107** (2010) doi:10.1073/pnas.1008157107 (Special Issue on Surface Chemistry).
7. Y. Han et al., Phys. Rev. Lett. **100**, 116105 (2008); Phys. Rev. B **81**, 115462 (2010).
8. M.C. Bartelt and J.W. Evans, Surf. Sci. Lett. **314**, L829 (1994).
9. G. Kresse and J. Hafner, Phys. Rev. B **47**, 558 (1993).
10. J.P. Perdew, K. Burke, and M. Ernzerhof, Phys. Rev. Lett. **77**, 3865 (1996).
11. G. Kresse and D. Joubert, Phys. Rev. B **59**, 1758 (1999).

Mater. Res. Soc. Symp. Proc. Vol. 1318 © 2011 Materials Research Society
DOI: 10.1557/opl.2011.626

In-situ TEM Observation of Formation-Retraction-Fracture Experiment of Liquid-Like Silicon Nanocontact

Tadashi Ishida[1], Kuniyuki Kakushima[1,2] and Hiroyuki Fujita[1]
[1] Center for International Research on Micronano Mechatronics, Institute of Industrial Science, University of Tokyo,
4-6-1 Komaba, Meguro, Tokyo, 153-8505, JAPAN.
[2] Tokyo Institute of Technology,
4259 Nagatsuda, Midori, Yokohama, Kanagawa, 226-8502, JAPAN.

ABSTRACT

We have examined the change of mechanical characteristics of silicon nanocontacts with and without the pre-treatment by flowing current through the contact. The silicon nanocontact formed between silicon tips by a mechanical contact was quickly deformed during its tensile test under a transmission electron microscope, after applying over 100 μA at a high bias voltage around 15 V between tips for a short duration. In the tensile experiment, the diameter of the nanocontact easily decreased from the initial diameter of 98 nm to 30 nm and the length increased from 11 nm to 66 nm. At 30 nm in diameter, it was suddenly fractured without further elongation and became round tips with smooth surfaces. According to the close observation, the silicon nanocontact seemed amorphous. In the retraction process of the silicon nanocontact, steps moved along the surface from the neck of the nanocontact to the tip side at the speed of 7.0 nm/s. Nano-scaled round step propagations were repeated from the neck to the tip. The step propagation caused the fast thinning of the nanocontact. On the other hand, the silicon nanocontact formed at 1 V in bias voltage was gradually thinned from 42.5 nm to 1.6 nm in diameter and elongated from 2.9 nm to 61.9 nm in length. From the comparison of silicon nanocontacts with and without the pre-treatment, the silicon nanocontact after flowing substantial current showed quick deformation and had different mechanical characteristics from the silicon nanocontact without the pre-treatment.

INTRODUCTION

Silicon is one of the most common materials for micro/nano devices [1,2]. It is widely used not only in very large scale integrated (VLSI) devices but also in micro electro mechanical systems (MEMS). The dimensions of the devices are minimizing to have better properties [3]. It is important to measure the electronic and mechanical properties of silicon at the nano scale because the properties are expected to be very different in such a small scale from ordinary scales. Many researchers have studied the electronic and mechanical properties of a silicon nanowire and a nanocontact for the further development of nano devices [4,5].

In the study on the electric and mechanical properties of nano structures, experiment with *in-situ* observation is favorable because the properties of nano structures are sensitively changed by their shapes [6]. Thus, *in-situ* observation was performed by a transmission electron microscope (TEM) with a help of small actuators, such as a piezoelectric actuator [7,8], a polymer membrane actuator driven by charge-up [9] and a MEMS actuator [10]. These equipments could achieve the electrical/mechanical measurement of nanostructures with *in-situ* observation. In the tensile experiment of silicon nanostructures, the different deformation behaviors were reported as follows. A silicon crystalline nanocontact was broken in a brittle behavior at 30 % strained position [11]. On the other hand, a silicon crystalline nanowire was elongated to twice as long as its initial length with changing its microstructure from crystal to amorphous [12]. A silicon nanocontact was also elongated to almost twenty times as long as the initial length in a ductile behavior at a quasi-static retraction speed [13]. In this article, we report that the silicon nanocontact with the pre-treatment of flowing the current over 100 μA for a short duration was quickly deformed and fractured during its tensile test, and keep its surface quite smooth.

EXPERIMENT

In our experimental system, MEMS device with micro electrostatic actuators were operated inside an ultra-high-vacuum TEM specimen chamber, 5×10^{-8} Pa. The schematic drawing of the MEMS device is shown in Fig. 1. It has a movable electrode, a fixed electrode, a pair of driving electrodes and opposing tips. The combination between the movable electrode and driving electrodes works as a micro electrostatic actuator. The micro electrostatic actuators are driven by the electrostatic force, with applying the driving voltage between the movable electrode and a pair of driving electrodes. The precision of the micro electrostatic actuator is better than sub-nm order and its stability is 0.1 nm in standard deviation for several tens of minutes.

The experimental setup composed of a MEMS device and a TEM is shown in Fig. 2. The MEMS device is inserted into a TEM specimen chamber using a specially-customized TEM sample holder. Thanks to the feed-throughs of this TEM holder, a driving voltage for the actuators and a bias voltage for an electrical measurement between tips are applied. The resolution of the current measurement is 500 pA in the full-scale of 100 μA. The current density of the electron beam for TEM observation is 1.6×10^{3} A/cm^2 without an appreciable heating according to a calculation [14]. Using this combination between the MEMS device and the TEM, a deformation of the nanostructure driven by the motion of MEMS tips can be *in-situ* visualized with TEM. The spatial resolution of TEM observation is 0.2 nm and it can be also recorded as a NTSC video at 30 fps.

Experimental procedure is shown in Fig. 3. The movable tip approaches the fixed tip while 1 V is applied between them (Fig. 3a). When opposing tips are brought into contact (Fig. 3b), a tiny electric current is detected. The bias voltage increases over ten volts, leading

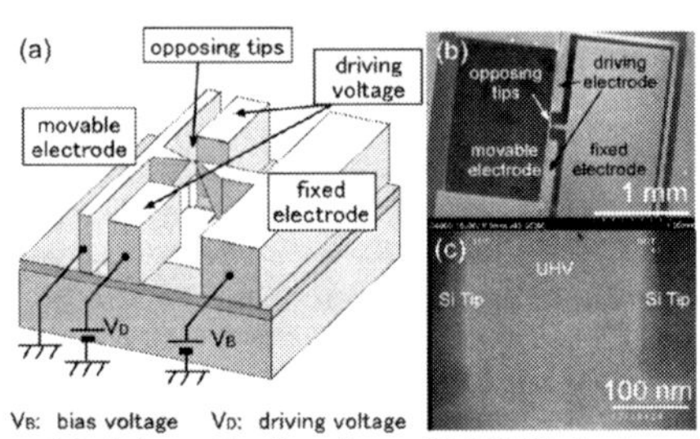

Figure 1. MEMS device. (a) Schematic drawing of MEMS device. (b) Overview of MEMS device. (c) Close-up image of opposing tips of MEMS device.

to a sudden jump in electrical current through a contact. At the moment, a silicon nanocontact is formed (Fig. 3c). After this sudden increase of the current, the bias voltage is decreased to 1 V to suppress the influence such as heating and electromigration. Tips are separated once and adjoined again (Fig. 3d). A silicon nanocontact appears between tips (Fig. 3e). The silicon nanocontact is retracted and finally fractured (Figs. 3f,g). All the processes are observed by the TEM in realtime. The shape of the silicon nanocontact is assumed as a hourglass.

RESULTS

Figure 4 shows the current response by the temporal change of the bias voltage after tips were brought into contact. The current was 13 nA at 1 V in bias voltage. The bias voltage gradually increased from 1 V to 15 V, while the driving voltage was kept constant. From 1 V to 13 V, no change was observed in the image (Fig. 5a). At 14 V, the current increased to 12 μA and the contrast of the tip around the contact point was changed (Fig.5b). This change seemed

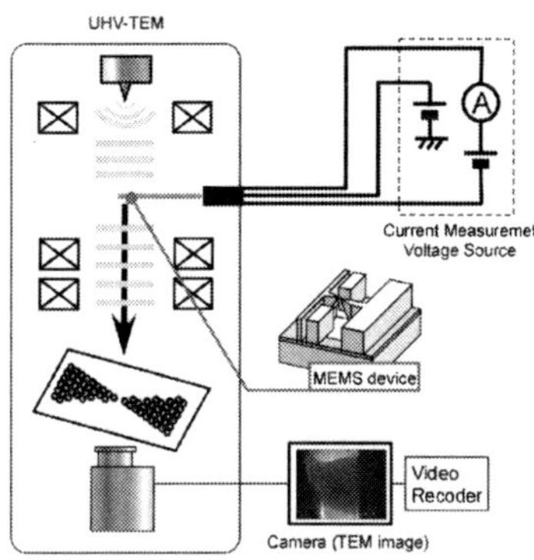

Figure 2. MEMS-in-TEM setup. MEMS device can be driven inside TEM specimen chamber. The electrical and mechanical properties can be measured between tips. The deformation was observed by a camera and recorded by a video recorder at 33 ms in temporal resolution.

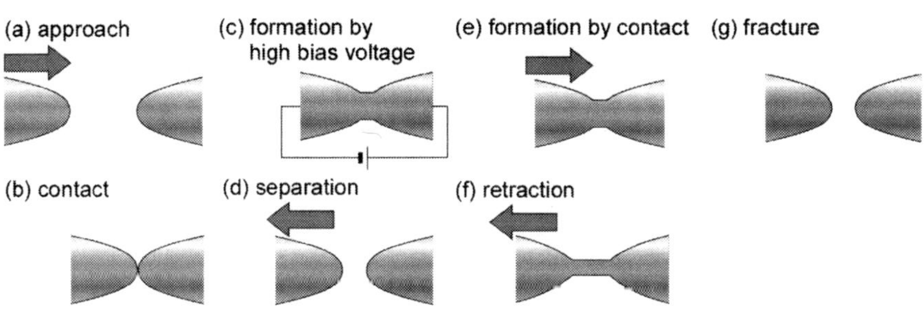

Figure 3. Experimental procedure of opposing tips. (a) Approach. (b) Contact. (c) Formation of nanocontact by a high bias voltage. (d) Tip separation. (e) Formation by mechanical contact. (f) Retraction. (g) Fracture.

the precursor to induce the contact formation at the higher voltage. At 15 V, a silicon nanocontact suddenly appeared between tips (Fig. 5c). Accordingly, electric current jumped from 12 μA up to 100 μA (Fig. 4). The bias voltage was set back to 1 V; the current decreased to 32 nA. Tips were separated from each other using an electrostatic actuator to identify the location of the silicon nanocontact. After this process, silicon tips were brought into slight contact, again, using the actuator. At the moment, a silicon nanocontact of 98 nm in diameter was instantaneously formed between tips. The silicon nanocontact did not show any lattice fringes with highly magnified observation; this suggests that it was an amorphous structure. The diameter of the necking drastically decreased from 98 nm to 30 nm with a

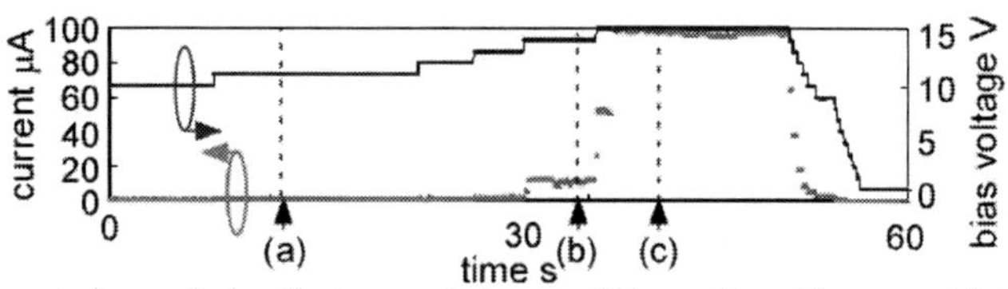

Figure 4. Current change during the temporal change of bias voltage. The current between silicon tips suddenly jumped up to 100 μA at 15 V in bias voltage. The insets from (a) to (c) corresponds to the insets in Fig. 5.

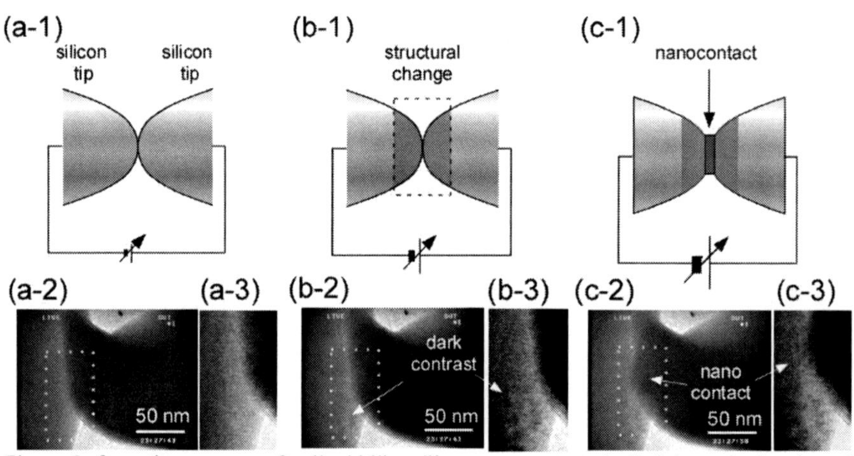

Figure 5. formation process of a liquid-like silicon nanocontact. (a-1) Schematic drawing of adjoined opposing tips with bias voltage of less than 14 V. (a-2) TEM image of adjoined opposing tips with bias voltage of less than 14 V. (a-3) high contrasted zoom-up image in dashed square of (a-2). (b-1) Schematic drawing of adjoined opposing tips with bias voltage of 14 V. Structural changes around the contact point. (b-2) Schematic drawing of tips at 14 V. Structural changes around the contact point in the yellow dashed square. (b-3) high contrasted zoom-up image in dashed square of (b-2). (c-1) Schematic drawing of formation of liquid-like silicon nanocontact at 15 V in bias voltage. (c-2) TEM image of liquid-like silicon nanocontact at 15 V in the yellow dashed square. (c-3) high contrasted zoom-up image in dashed square of (c-2).

retraction of 55 nm in 22 seconds (Fig. 6). The silicon nanocontact was elongated from 11 nm to 66 nm. During this deformation, nano scaled round steps of 4.4 nm in height propagated along the surface from the neck to the tip at 7.0 nm/s (Fig. 6b,c). The nanocontact was fractured without any further retraction, when the diameter and the length of the silicon nanocontact were 30 nm and 66 nm, respectively. The corner of the breakage point wasrounded to 12 nm in radius of curvature within 33 ms. The silicon nanocontact kept this behavior, when the formation-retraction-fracture process was repeated for 1 hour.

DISCUSSION

The similar behavior of the silicon nanocontact was observed for the nanocontact after the current suddenly jumped up to 100 μA at high bias voltages in 16 experiments. The threshold voltage to induce the change was not always the same. It varied from 13 V to 20 V. With the pre-treatment of flowing 100 μA, the silicon nanocontact was suddenly formed when the tips were brought into contact. It was broken when the diameter was a few tens of nanometers. The broken corner was rounded to a few tens of nanometers within 33 ms; its radius curvature did not increase anymore.

On the other hand, without the pre-treatment, the silicon nanocontact was formed by the

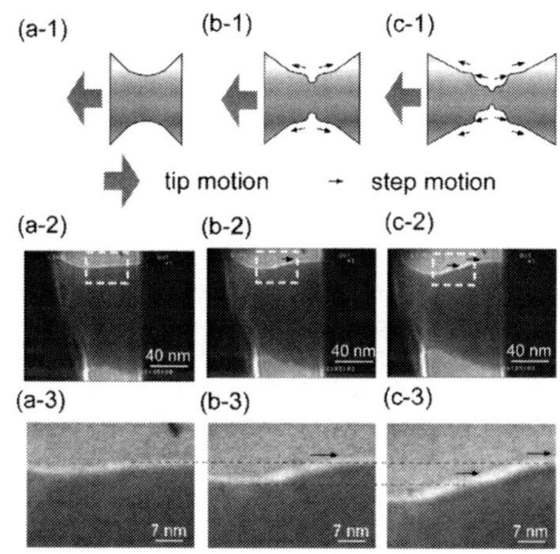

Figure 6 Retraction process of a silicon nanocontact with the pre-treatment. (a-1) schematic drawing of silicon nanocontact during retraction process. (a-2) TEM image of silicon nanocontact during retraction process. (a-3) zoom-up image of silicon nanocontact in yellow dashed square in (a-2). (b-1) Schematic drawing of round step formation on surface of silicon nanocontact due to back motion from (a-1). (b-2) TEM image of step formation on silicon nanocontact surface due to back motion of 17 nm from (a-2). (b-3) zoom-up image of silicon nanocontact in yellow dashed square in (b-2) which tracked the place of yellow dashed square in (a-2). (c-1) Schematic drawing of another step formation on silicon nanocontact due to back motion from (b-1). (c-2) TEM image of step formation on silicon nanocontact due to back motion of 15 nm from (b-2). (c-3) zoom-up image of silicon nanocontact in yellow dashed square in (c-2) which tracked the place of yellow dashed square in (a-2, b-2).

gradual growth for a few tens of minutes [15]. The nanocontact was retracted from a few nm to several tens of nanometers. It was broken at 1.6 nm in diameter and the broken corner was rounded from sub-nm order to over 1 nm in a few minutes [16]. By comparing silicon nanocontacts with and without the pre-treatment, we felt the former behaved like a liquid because it deforms so easily and quickly.

CONCLUSIONS

The *in-situ* observation of the tensile test of silicon naonocontacts was achieved using the combination between the MEMS device and the TEM. The deformation of a silicon nanocontact with pre-treatment of flowing 100 μA was completely different from the silicon nanocontact without such pre-treatment. This result suggested that the properties of nanostructures can be easily affected by the fabrication process and treatment. It is necessary to know the fabrication history of nano-scaled devices even if the devices seemed similar in their shapes.

ACKNOWLEDGMENTS

This work was supported by KAKENHI; the grant-in-aid of Scientific Research (S) 16191004 and Specially Promoted Research 21000008 sponsored by Japan Society for the Promotion of Science (JSPS).

REFERENCES

1. M. Leong, B. Doris, J. Kedzierski, K. Rim and M. Yang, *Science*, **306**, 2057 (2004).
2. D. F. Moore and R. R. A. Syms, *Electronics & Communication Engineering Journal*, **11**, 261 (1999).
3. G. Moore, Electronics, **38** (1965).
4. Y. Cui, Z. Zhong, D. Wang, W. U. Wang and C. M. Lieber, *Nano Lett.* **3**, 149 (2003).
5. S. Hoffmann, I. Utke, B. Moser, J. Michler, S. H. Christiansen, V. Schmidt, S. Senz, P. Wermer, U Gosele and C. Ballif, *Nano Lett.* **6**, 622 (2006).
6. Y. Oshima, K. Mouri, H. Hirayama and K. Takayanagi, *J. Phys. Soc. Jpn.* **75**, 053705 (2006).
7. T. Kizuka, *Phys. Rev. B*, **63**, 033309 (2001).
8. Y. Oshima, K. Mouri, H. Hirayama and K. Takayanagi, *Surf. Sci.* **531**, 209 (2003).
9. T. Kizuka, H. Ohmi, T. Sumi, K. Kumazawa, S. Deguchi, M. Naruse, S. Fujisawa, S. Sasaki, A. Yabe and Y. Enomoto, *Jpn. J. Appl. Phys.* **40** L170 (2001).
10. X. D. Han, Y. F. Zhang, K. Zheng, Z. Zhang, Y. J. Hao, X. Y. Guo,J. Yuan, Z. L. Wang, *Nano Lett.* **7**, 452 (2007).
11. T. Ishida, Y. Nakajima, K. Kakushima, M. Mita, H. Toshiyoshi and H. Fujita, *J. Micromech. Microeng.* **20,** 075011 (2010).
12. T. Kizuka, Y. Takatani, K. Asaka and T. Yoshizaki, *Phys. Rev. B*, 72 035333 (2005).
13. X. Han, K. Zheng, Y.-F. Zhang, X. Zhang, Z. Zhang and Z.-L. Wang, *Adv. Mater.* **19**, 2112 (2007).
14. S. Iijima, *Microclusters* ed. Sugano S, Nishina Y and Ohnishi S (Springer-Verlag) pp. 186-192 (1987).
15. T. Ishida, K. Kakushima, N. Sasaki and H. Fujita, *Nanotech.* **21,** 435705 (2010).
16. T. Ishida, K. Kakushima, M. Mita, H. Fujita, *Technical Digest of MEMS 2005*, 879 (2005).

Mater. Res. Soc. Symp. Proc. Vol. 1318 © 2011 Materials Research Society
DOI: 10.1557/opl.2011.623

Observation Of Real-Time Thin Film Evolution Using Microcantilever Sensors

Alan M. Schilowitz and Dalia G. Yablon
Corporate Strategic Research, ExxonMobil Research and Engineering
Annandale, NJ 08801

ABSTRACT

Adsorption and desorption kinetics of thin film formation on metal surfaces has been directly monitored in real-time by optically measuring the deflection of activated atomic force microscope microcantilevers. Microcantilever deflection is caused by stress generated during the formation of an adsorbate layer on one side of the microcantilever. In this work, rapid adsorption of carboxylic acid in hydrocarbon solvent onto a gold surface was directly observed as a compressive stress developed on the microcantilever substrate. Upon exposure to alkylthiol, acid desorbed and was displaced by alkylthiol; this process was continuously monitored in real-time. Experiments suggest that some film organization and intermolecular interaction are required before substantial surface stress can be detected.

INTRODUCTION

Microcantilevers (MC) developed for atomic force microscopes have been shown to be useful tools for sensitive sensing of adsorption phenomena [1,2]. MC sensors are generally operated in either dynamic or static mode. In dynamic mode, cantilevers can detect mass change and viscous damping by monitoring changes that occur in the cantilever's vibrational resonance spectrum. Adsorption induced resonance changes have also been observed that are due to changes in the cantilever's force constant [1,3,4]. Static mode cantilevers coated on one side with an activating layer such as gold bend due to surface stress caused by preferential adsorption [5,6,7,8]. These applications of MC sensors all depend on selective adsorption. The impact of competing adsorption processes on MC sensor deflection is a relevant issue that has not had much attention and is especially relevant to sensing in multicomponent systems or cases where the cantilevers become contaminated before use. Cantilever deflection affords sensitive, real-time, *in situ* observation of competing adsorption/desorption phenomena in real-time.

Our interest in MC sensors derives from the potential for using these devices in static mode for detecting contaminants that might occur in hydrocarbons. Small mercaptans (e.g. sulfur containing molecules with less than 10 carbons) and carboxylic acids are potential species of interest. Application of such sensors will require them to function in environments where significant surface contamination will be impossible to prevent. The effect of pre-contamination to improve sensor selectivity is discussed. Calibration of MC deflection with concentration is also discussed. Finally, the use of MC sensors to monitor real-time adsorption and desorption phenomena is demonstrated

EXPERIMENTAL

Silicon contact mode microcantilevers (Budget Sensors) with a nominal resonance frequency of 13 kHz and a nominal force constant of 0.2 N/m were used in all experiments. A 25 nm thick gold layer was deposited onto one side of the cantilever in a vapor deposition chamber.

The cantilevers were used without further cleaning. Deflection experiments were started approximately four hours after cantilevers were coated with gold.

Oleic acid (Aldrich 36,452-5), n-heptane (JT Baker), 1-octadecene (Aldrich O-80-6) and 1-nonanethiol (Aldrich N3,140-0) were used without additional purification. Cantilever motion was monitored using the optical beam deflection method [1]. Cantilever deflection was monitored by amplifying the output of a one-dimensional Position Sensitive Detector (PSD) (Hamamatsu) and acquiring data at a rate of six points per minute. The cantilever was held in a purpose built thermally insulated stainless steel flow cell with a liquid volume of 3.5 ml. A one-inch diameter quartz window was secured to the top of the cell.

Liquids were introduced into the cell using syringes mounted in a syringe pump. Flow rate was maintained at 2 microliters per minute, except when the solution was changed in which case a higher flow rate was used until sufficient fluid had flowed through the cell to exchange its contents several times. Throughout the experiments temperature was measured at 23.3 C +/- 0.5. However, within a single day's experiments temperature variation was +/- 0.25 C. Over this temperature range deflection due to dissimilar thermal expansion of the cantilever and its gold coating is relatively small [6].

Depending on experimental conditions, cantilevers can deflect in either direction. We will refer to a deflection causing compression of the cantilever by the gold surface as an upward deflection in the positive direction.

DISCUSSION

A cantilever coated on one side with gold was mounted in the flow cell and immersed in heptane. After three hours, temperature fluctuations dissipated and the cantilever achieved a relatively stable baseline. Cantilever deflection caused by acid adsorption was explored by injecting a 0.001 M solution of oleic acid dissolved in n-heptane. Adsorption of oleic acid resulted in rapid upward cantilever deflection without a significant induction period as shown in Figure 1. This experiment was repeated with a stearic acid solution with a similar deflection result. The deflection response shown in Figure 1 is similar but faster to that observed with mercaptans on gold coated microcantilevers - under similar conditions - even though carboxylic acids do not chemisorb and form stable Self-Assembled Monolayers (SAM) on gold. The nature of acid adsorption was explored further by flushing the cell and cantilever with n-heptane followed by injection of a 0.003 M solution of 1-nonanethiol dissolved in n-heptane. The thiol solution induced a downward cantilever deflection, in contrast to the upward deflection previously induced by the acid solution. After the initial downward deflection, a slow upward deflection takes place. This demonstrates the use of cantilever deflection as a means for real-time observation of adsorption and desorption phenomena

It appears that the reversal of deflection evident in Figure 2 at approximately 50 minutes is representative of acid desorption from the cantilever surface. Cantilever redeflection at approximately 60 minutes represents replacement of acid by thiol which binds more strongly to the gold surface than the acid. This explanation is further supported by the fact that the thiol induced deflection in Figure 2 occurs at about the same rate as deflection caused by direct injection of thiol solution without first applying oleic acid. Reinjection of oleic acid after 1-nonanethiol did not cause desorption of the more strongly bound thiol. Figure 2 also contains a best fit curve of the experimental data points to a double exponential function representing a consecutive A-B-C reaction where the initial reactant is oleic acid bound to the gold surface, the intermediate product is a clean gold surface and the final product is a gold surface covered with thiol.

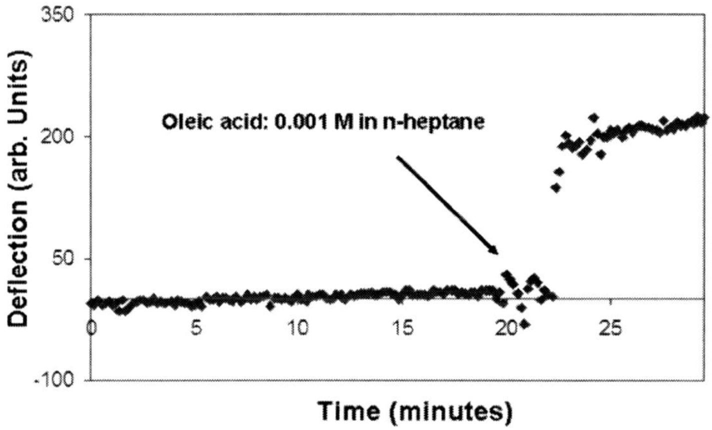

Figure 1. Deflection time trace of freshly coated cantilever coated on one side with gold that has been stabilized in n-heptane. A solution of 0.001 M Oleic acid was injected as indicated by the arrow. Note that increasing deflection in these figures represents a compressive deflection of the cantilever. Deflection units are arbitrary.

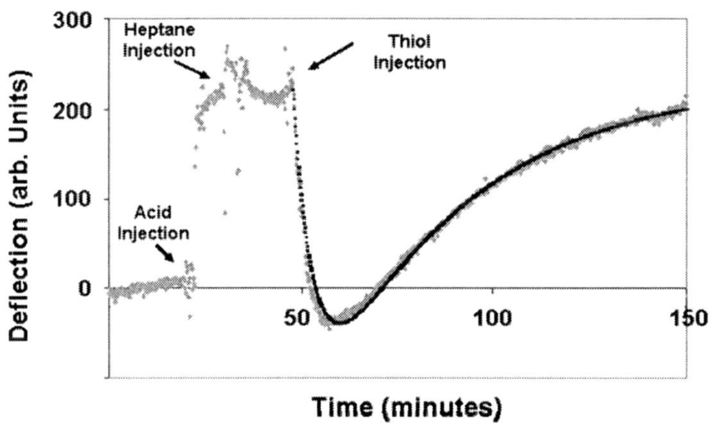

Figure 2. Deflection time trace of freshly coated cantilever coated on one side with gold that has been stabilized in n-heptane. Oleic acid (0.001 M), n-heptane, and 1-nonanethiol (0.003 M) solutions were injected at various times as indicated by arrows. Solid black line is double exponential fit to data points. Deflection units are arbitrary.

The time trace in Figure 2 indicates that acid desorption occurs at a significantly faster rate than adsorption of thiol. This may be due to the fact that deflection is directly related to surface stress and is only an indirect measure of adsorption and desorption. The kinetics of adsorption may differ from the kinetics of surface stress. Alternatively, the binding density of alkyl thiols, which can form dense monolayers on gold, is different than that of carboxylic acids, which form a less dense lying down phase.

While adsorption of oleic acid to the silica side of the cantilever is likely, the shape of the time trace in Figure 2 after about 50 minutes is due to displacement of acid by thiol from the gold surface. This suggests that stress induced cantilever deflection caused by oleic acid is related to physisorption of acid to the gold side of the cantilever. It was further confirmed that physisorption could induce cantilever deflection by ensuring that a similarly gold coated cantilever deflected upon exposure to a 0.001 M solution of 1-octadecene dissolved in n-heptane. Alkenes are known to physisorb to gold [9]. While 1-octadecene did cause deflection, the stress-induced deflection was significantly less than the deflection caused by oleic acid, presumably due to the ability of oleic acid to form a more stable hydrogen bonded network on the surface which induces greater compressive stress on the cantilever.

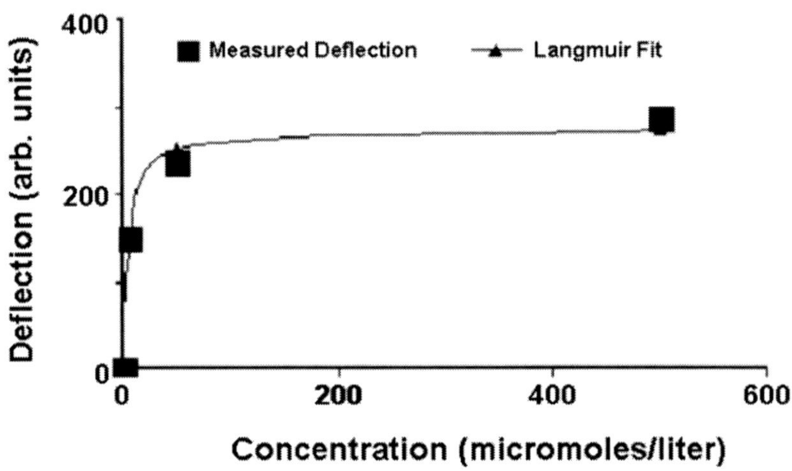

Figure 3. Deflection response of a single microcantilever after exposure to different concentrations of 1-nonanethiol dissolved in n-heptane. Each point represents the maximum deflection reached by the microcantilever at each concentration. Deflection level is in arbitrary units. The solid line represents the best fit of a Langmuir isotherm to the data points.

The utility of microcantilevers for practical sensing applications is dependent on calibrating deflection to concentration. This is straightforward if deflection can be related to a simple isotherm such as the Langmuir adsorption isotherm. If deflection is affected by a confounding factor such as evolving surface stress (e.g. maturation of an ordered surface film) this would complicate quantification and potentially increase the time needed to make a measurement. Concentration dependent deflection of 1-nonanethiol solutions was measured to calibrate deflection. These experiments were carried out with a single microcantilever by sequentially exposing it to solutions with increasing concentration of 1-nonanethiol dissolved in n-heptane. In each case a solution was injected into the cell and increased cantilever deflection was measured. Each successive solution was injected after about one hour when the cantilever deflection reached a steady state level. Figure 3 shows a plot of relative deflection vs. concentration and the best fit Langmuir isotherm is plotted through the experimental points. Application of Langmuir behavior assumes that deflection is directly proportional to adsorbate concentration and surface coverage.

The Langmuir fit in Figure 3 suggests that deflection is easily related to concentration and the sensor can be readily calibrated without much complication. However, the Langmuir fit yields a free energy of adsorption of -7.2 kcal/mol which is more negative than reported for other thiols in other solvents. For example the free energy for adsorption of octadecanethiol dissolved in n-hexane onto a gold surface is reported as -5.72 kcal/mol at 298° K [10]. This discrepancy could be due to inherent chemical differences in the systems or experimental error. More likely it suggests a partial breakdown in the assumption that deflection is directly related to concentration over the full range of concentration explored here. This might be attributed to several factors even if the adsorption itself follows Langmuir behavior. For example, the MC spring may not behave linearly over the full range explored here. Alternatively, stress might not be directly proportional to surface coverage.

CONCLUSION

Cantilever deflection is a useful methodology for real-time studies of kinetics of surface phenomena where adsorption onto an activated surface leads to increased or decreased surface stress. Competitive adsorption of 1-nonanethiol dissolved in n-heptane onto a gold-coated Si cantilever pretreated with oleic acid was explored by monitoring microcantilever deflection. Upon introduction of thiol, cantilever bending is characterized by relatively fast desorption of the oleic acid followed by slower adsorption of the thiol. Studies of competing adsorption processes using gold-coated cantilevers are relevant to application of such sensors in multicomponent systems and applications where cantilevers might become contaminated before use. For example, chemical specificity can be imparted to an array of cantilever sensors by pre-treatment of cantilever surfaces with adsorbates of different binding affinities. Alternatively, pre-treatment of cantilever surfaces might be used to protect or preserve a cantilever from environmental contaminants. Interestingly, oleic acid caused a relatively large and fast deflection response which is attributed to formation of a hydrogen bonded network on the gold surface.

The relationship between cantilever deflection and concentration was also explored. Results show that, in liquid phase, adsorption of thiol generally follows Langmuir behavior as originally suggested in Reference 6 for vapor phase experiments. This enables consistent calibration of the sensor over a relatively large concentration range.

REFERENCES

[1] Thundat,T.; Oden, P.I.; Warmack, R.J. *Micro. Thermophys. Eng.* **1997**,*1*, 185-199.
[2] Wachter, E.A.; Thundat, T. *Rev. Sci. Instrum.* **1995**, *66*, 3662-3667.

[3] Khaled,A.R.A.; Vafai, K.; Yang, M.; Zhang, X.; Ozkan, C.S. *Sens. Actuators, B, Chem.* **2003**, *94*, 103-115.
[4] Cherian, S; Thundat, T. *Appl. Phys. Lett.* **2002**, *80,* 2219-2221.
[5] O'Shea, S.J.; Welland, M.E.; Brunt, T.A.; Ramadan, A.R.; Rayment, T. *J. Vac. Sci. Technol. B*, **1996**, *14*, 1383-1385.
[6] Berger, R.; Delamarche, E.; Lang, H.P.; Gerber, Ch.; Gimzewski, J.K.; Meyer, E.; Guntherodt, H.J. *Science*, **1997**, *276* , 2021-2024.
[7] Berger, R.; Delamarche, E.; Lang, H.P.; Gerber, Ch.; Gimzewski, J.K.; Meyer, E.; Guntherodt, H.J. *Appl. Phys. A*, **1998**, *66*, S55-S59.
[8] Datskos, P.G.; Sauers, I. *Sens. Actuators, B, Chem*, **1999**, *61*, 75-82.
[9] Wetterer, S.M.; Lavrich, D.J.; Cummings, T.; Bernasek, S.L.; Scoles, G. *J. Phys. Chem. B* **1998**, *102*, 9266-9275.
[10] Karpovich, D.S., Schessler, H.M., Blanchard, G.J., Thin Films, Academic Press NY, **1998**, *24,* 44-81.

Mater. Res. Soc. Symp. Proc. Vol. 1318 © 2011 Materials Research Society
DOI: 10.1557/opl.2011.624

X-Ray Study of Strained and Strain Balanced Superlattice Materials

Natee Johnson[1], Ruth Choa[3], Liwei Cheng[2], Fow-Sen Choa[2]
[1]Mechanical Engineering, University of Maryland, Baltimore County, Baltimore, MD, USA.
[2]Computer Science and Electrical Engineering, University of Maryland, Baltimore County, Baltimore, MD, USA.
[3]School of Engineering and Applied Science, Harvard University, Cambridge, MA, USA.

ABSTRACT

Nano-scale superlattice (SL) based devices, such as quantum cascade lasers QCLs, have recently become very important due to their capability to identify toxic and explosive chemicals. In manufacturing these Mid-IR photonic devices, atomic-level scanning tunneling microscopes (STM) and transmission electron microscopes (TEM) have been used to characterize the growth quality of superlattice wafers. However, these methods yield observations that are localized and cannot view the entire structure and even now we have not been able to correlate these measured crystal lattice images with device performance. The x-ray scanning technique has greater likelihood of success given that it can observe not only the localized, but also the entire superlattice structure. By extracting special features and key parameters in x-ray diffraction (XRD) patterns, the epitaxial quality of QCL superlattices can be evaluated and correlated to the performance of fabricated QCL devices. We can then differentiate and classify different grades of wafers before starting device fabrication and testing.

Such an example of the usefulness of XRD can be found with strain-balanced superlattices, such as InGaAs/InAlAs, where there is notable decrease in laser performance after relaxation. It can also be found with type-II InAs/GaSb strained layer superlattice, which is currently the best candidate for room temperature mid-IR detectors and focal plane arrays. Experimentally measured x-ray patterns are compared to simulation results and problem sources are identified.

INTRODUCTION

Nano-scale SL based devices such as quantum well infrared photodetectors (QWIPs) and QCLs have existed for more than a decade. However, even now, we still lack of a systematic methodology to study and understand the relationship between material quality and device performance. For example, interface and surface states affect interband devices significantly because their carrier lifetime is in the nanosecond to microsecond scale. On the other hand, it is not clear how hetero-interface states affect the performance of intersubband devices because their lifetime (intersubband state) is in the picosecond to sub-picosecond level. Furthermore, a typical 30-period QCL can easily have >1000 slices of nano-scale-thick layers with a total thickness of a few microns. The analyzing method needs to be able to simultaneously examine the quality of all the layers in each one of the 30 periods.

The diffraction of an x-ray through a superlattice material is a Fourier transformation of the superlattice structure (in x-space) to the k-space, expressed by the profile of diffraction angles, where k is the wave propagation constant or the particle momentum [1]. By extracting key features in the Fourier domain, we can explain as well as predict the corresponding layer thickness and chemical composition variations in the x-space. By using the special features and key parameters in x-ray diffraction patterns, the epitaxial quality of QCL superlattices can be evaluated and we can differentiate and classify different grades of wafers before starting device fabrication and testing. This saves processing time (at least 3-4 weeks) because poor wafers can be immediately eliminated, and performance for all other wafers can be anticipated prior to fabrication. Most importantly, the fast feedback helps crystal growers improve the next growth run without weeks of waiting.

EXPERIMENT

Fourier analysis can be used to analyze both time and spatial domain signals. Usually it is difficult to directly characterize signals in the time or spatial domain. However, when these signals are transformed to the Fourier (frequency- or k-) domain, the features and basic characteristics of can be easily identified. By quantitatively measuring important parameters in the Fourier domain, we can effectively evaluate the quality and identify key features of the object. Simulations were done on x-ray scan profiles obtained from many metal organic chemical vapor deposition (MOCVD) and molecular beam epitaxy (MBE) grown superlattice wafers. We started with very simple structures and then step-by-step interpreted the profiles corresponding to lattice mismatching, interfacial defects, and degradation of periodicity or symmetry caused by growth rate changes, chemical composition changes, lattice relaxation of very complex QCL structures. The BEDE simulation program is used to match theoretical x-ray diffraction patterns with experimental results. Parameters analysed include material layer thickness, composition, hetero-interfaces, strain and relaxation. The measured characteristics of fabricated lasers and detectors will be compared to the predictions based on the statistical results from the data fitting and similarity-extraction. This will allow device performance to be correlated with x-ray scan results and serve as a means to verify the accuracy of our predictions. The following show some examples and their interpretations.

DISCUSSION

Since strain-balanced superlattices are composed of alternating compressive and tensile strained layers, the thicker layers are sometimes strained too much and lattice relaxation occurs - chemical bonds between atoms break and lead to serious degradation of periodicity. A very small percentage of relaxation can lead to a dramatic broadening of satellite peaks and a serious decrease in laser performance. Figure 1(a) shows the x-ray scan result of a strained superlattice sample with serious lattice relaxation. Nearly all satellite peaks are broadened. Figure 1(b) shows the simulated result with 200-period of InGaAs/InAlAs superlattice and ~2% relaxation in some of the thicker InAlAs layers. Through simulations, relaxations in InGaAs and InAlAs layers were found to show distinct patterns, which helps in determining which exact materials are causing the problems in a superlattice. As shown in Figure 2, relaxation in InGaAs layers causes each peak to tilt upwards from the left to the right and relaxation in InAlAs causes individual peaks to tilt downwards from the left to the right.

Simulations and experimental results for the 100 period InAs/GaSb superlattice are compared as a means of analysis but are not yet conclusive. Preliminary matching results are shown in Figure 3 where the red experimental line requires further manipulation to adequately depict the parameters of the sample.

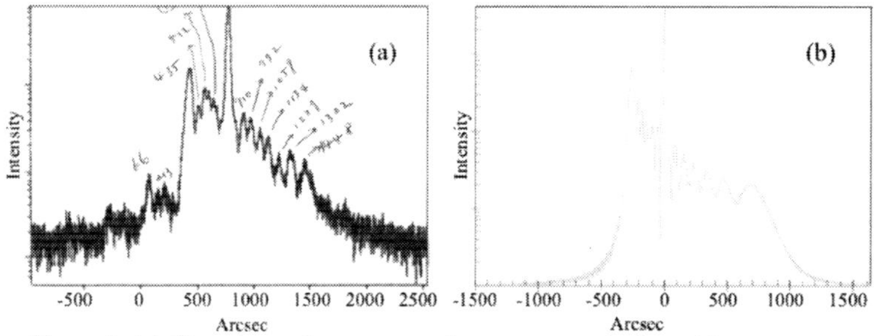

Figure 1. (a) Experimental x-ray scan of a grown superlattice with severe relaxation – we can identify that the problems are occurring in the layers of InAlAs because of the way the peaks tilt. **(b)** Simulated scan results of a super lattice with ~2% lattice relaxation taking place in some of the thicker InAlAs layers.

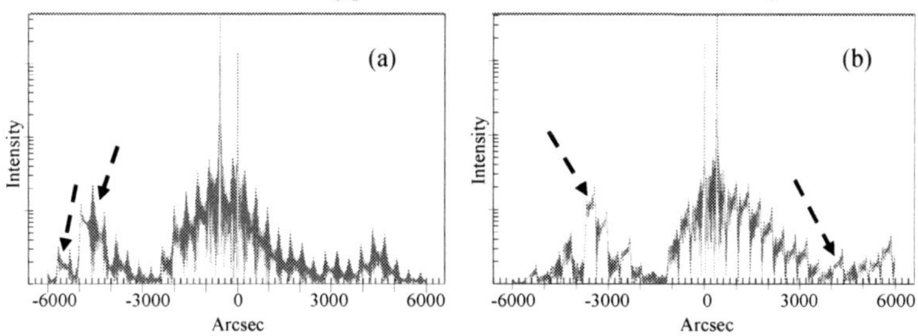

Figure 2. (a) 0.5% relaxation in the thicker layers of the InAlAs material – note how the peaks tilt downwards from the left to the right. **(b)** 0.5% relaxation in the thicker layers of the InGaAs layers – note how the peaks tilt upwards from the left to the right, as indicated by black arrows.

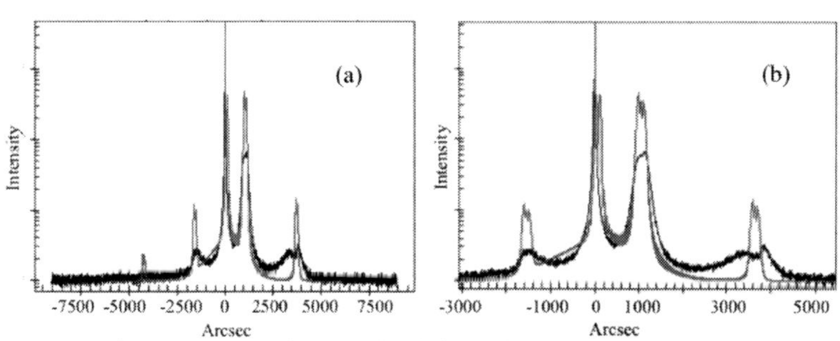

Figure 3. (a) and **(b)** Preliminary simulations of 100-period InAs/GaSb. Black curve represents experimental data from x-ray scans. Red curve represents simulated fitting from BEDE software.

CONCLUSION

For QCLs, hopefully more data on QCL performance measurement can be collected, statistically correlated with x-ray scan spectra results, and corresponded with physical implications and interpretations. It is hoped that the methods developed in this research can also be applied to other types of quantum devices such as quantum well infrared photodetectors (QWIPs) [2] and strained-layer superlattice (SLS) detectors [2].

REFERENCES

[1] David Keith Bowen, Brian Keith Tanner, High resolution X-ray diffractometry and topography, CRC Press, 1998.

[2] Roberto Paiella, Intersubband Transitions in Quantum Structures, McGraw-Hill Companies, January, 2006.

Mater. Res. Soc. Symp. Proc. Vol. 1318 © 2011 Materials Research Society
DOI: 10.1557/opl.2011.268

In-situ XRD and FIB microscopy studies of the dynamics of intermetallic phase formation in thin layer Cu/Sn films for low-temperature isothermal diffusion soldering

Harald Etschmaier[1,2], Jiří Novák[2], Hannes Eder[1], Peter Hadley[2]
[1]Unit Process Development, Infineon Technologies Austria AG
Villach 9500, Siemensstrasse 2, Austria
[2]Institute of Solid State Physics, Graz University of Technology
8010, Petersgasse 16, Austria

ABSTRACT

In this contribution we report on the dynamics of the phase evolution in electrochemically deposited Sn thin films on copper coated substrates studied by in-situ X-ray diffraction (XRD) and Focused Ion Beam Microscopy (FIB). The data obtained is used to extract fundamental parameters such as the activation energy and the rate constant of the reaction. Results indicate that the formation of intermetallic phases in these thin layers, in which the grain size exceeds the layer thickness, is not limited by diffusion but rather by reaction kinetics.

INTRODUCTION

Isothermal diffusion soldering (IDS) is a promising method for forming mechanically and thermally stable Pb-free bonds in electronic devices[1]. It combines good joint filling in large area joints and tolerance to surface roughness of conventional soldering with better control of joint thickness and reduced thermal expansion mismatch stress that is obtained from diffusion bonding[2]. For diffusion soldering a thin layer of a low melting temperature component like tin is deposited between two high melting temperature joint components like copper. Upon heating to bonding temperature at first the component with the lower melting point is melted. It quickly reacts with the high melting component to form an intermetallic phase whose melting point is substantially higher than the bonding temperature. In this way highly stable joints are achieved under relatively mild process conditions. The intermetallic phase formation in IDS has so far only been studied ex-situ mainly by optical microscopy, scanning electron microscopy, energy dispersive X-ray analysis and transmission electron microcopy of joint cross-sections or X-ray diffraction of fractured joints[3-9]. The drawback of these methods is that they can only give insight in the composition of the quenched sample, and the intermetallic phase evolution at process temperature cannot be observed directly. Furthermore, the investigated volume of these methods is very small and not representative of the whole sample. Therefore an in-situ, real time analysis of a macroscopic volume is necessary to trace the course of the reaction.

EXPERIMENT

Thin metal films were deposited on silver coated silicon wafers in an industrial scale plating line. A 5 μm thick film of copper was electrodeposited on the silver surface from industrial electrolytes at room temperature and a current density of 3 A/dm^2. This corresponds to a deposition rate of 0.66 μm/min. One μm of tin was deposited at a current density of 2 A/dm² and at rate of 1 μm/min on top of the copper.

For the in-situ XRD measurements a Bruker D8 Discovery diffractometer with Cu Kα radiation was used. The primary beam was collimated and monochromatized with a parabolic graded multilayer mirror (Göbel mirror) and a slit. The primary beam size was 0.1 mm with a divergence of 0.04 degrees. Diffraction patterns were recorded by a linear position sensitive detector VANTEC-1, enabling the capture of an angular range of $\Delta(2\theta)\approx12°$ in one frame. The detector's angular resolution was approximately 0.04 degrees. The sample heating was performed in an Anton Paar DHS 1100 heating chamber covered with a hemispherical PEEK dome and filled with He or Ar protective atmosphere. The samples were annealed isothermally at 5 different temperatures (120°C, 150°C, 180°C, 200°C, 350°C) following a rapid ramp up. At 350°C the 2Θ scattering angles ranging from 51° to 63° were collected in one shot enabling fast time-resolved in-situ annealing experiments. For slower processes at lower temperatures, a 2Θ range from 35° to 70° was recorded every 5 minutes. The qualitative phase analysis was performed using data from the Powder Diffraction File (International Center for Diffraction Data) and the Inorganic Crystal Structure Database (Fachinformationszentrum Karlsruhe).

The fiber texture of the electrodeposited copper and tin layers was examined by XRD pole figure analysis. The data was obtained using a Philips X'Pert System with Cr Kα radiation and a graphite monochromator on the secondary side. The diffractometer was equipped with an Eulerian cradle (four-circle goniometer) to allow tilting and rotating the specimen as required for texture measurements. The defocusing of the pole figures was corrected via powder normals.

Images of layer stack cross sections were taken in a FEI Vectra 200 focused ion beam microscope. The FIB was chosen over an electron microscope for its superior phase contrast. For cross-sectional imaging the silicon wafer substrate was broken along a preferred orientation to produce a smooth edge. Thereafter the surface of the obtained cross section was polished with a gallium ion beam at an acceleration voltage of 30 kV and a beam current of 20 nA. This way the image could be taken perpendicular to the surface without distortions. The acceleration voltage of the gallium ion beam used for imaging was also 30 kV with a beam current of 9 pA.

RESULTS AND DISCUSSION

Owing to the anisotropic property variations observed in copper and tin the film textures were of importance for this study[10]. By recording pole figures for multiple reflexes a strong <112> texture of tin was determined, whereas the copper grains are oriented randomly in the film.

According to the equilibrium phase diagram[11] there are two intermetallic phases of different compositions that are stable at the assayed temperatures: the copper rich ε-phase (Cu_3Sn) and the tin rich η-phase (Cu_6Sn_5). Upon heating of the Cu/Sn thin film couples the two metals react forming these two intermetallic phases. Figure 1 shows FIB micrographs of the evolving Cu/Sn interface upon annealing at 180°C. Initially the two metals are cleanly separated. During annealing the intermetallics grow in a layered structure in which the copper rich ε–phase forms adjacent to the copper and the tin rich η–phase adjacent to the tin. After 300 min most of the tin has been consumed and both intermetallic phases are present in similar abundances. By further annealing the Cu_6Sn_5 is transformed into Cu_3Sn which is the final state after prolonged annealing.

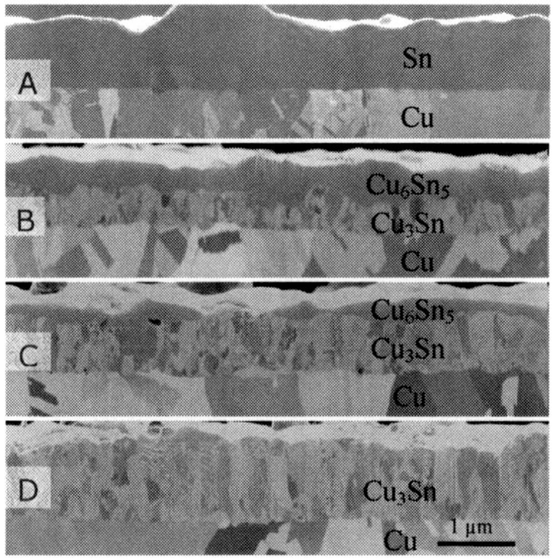

Figure 1: FIB micrographs of the Cu/Sn thin film couples. (A) After electrodeposition. (B) After annealing at 180°C for 300 min. (C) After annealing at 180°C for 600 min. (D) After annealing at 180°C for 1000 min. The intermetallic phases grow in a layered structure in which the copper rich ε–phase forms adjacent to the copper and the tin rich η–phase adjacent to the tin.

This process could be verified by in-situ XRD. Figure 2 shows the time evolution of the relative abundances of the intermetallic phases as measured by XRD at annealing temperatures of 180°C and 350°C. Similar plots have been generated for temperatures of 120°C, 150°C and 200°C but are not shown in this contribution. The data was obtained from integrating the intensities of several XRD peaks associated with each phase and normalizing to equilibrium intensity at the end of the annealing process. At the 350°C anneal which is well above the melting temperature of tin of 232°C a difference in the phase composition compared to lower temperature anneals was found. XRD peaks at 2Θ positions of 42.8° and 62.3° indicate the formation of a CuSn phase [12]. It is assumed that this is a metastable phase that forms temporarily at high temperatures. Its evolution is closely related to the stoichiometrically similar Cu_6Sn_5.

It was found that at all assayed temperatures Cu_3Sn grows in a linear fashion independent of the abundance of tin or Cu_6Sn_5. The linear growth, which has also been observed before by Rutherford backscattering[5], is in contradiction to the standard diffusion limited model. This predicts a square root time dependence, as it has been found for thick films[9]. Since marker results indicate that copper is the dominant diffusing species[13] the growth of Cu_3Sn is taking place at the Cu_3Sn/Cu_6Sn_5 interface. The growth requires the supply of copper atoms from the bulk copper and their diffusion through the Cu_3Sn layer to the Cu_3Sn/Cu_6Sn_5 interface where they react with Cu_6Sn_5. Since the growth speed of Cu_3Sn is constant even if a fraction of the copper atoms is used to form Cu_6Sn_5, the release of copper atoms cannot be the rate limiting factor. Therefore it can be assumed that the reaction of copper with Cu_6Sn_5 is the rate limiting step for Cu_3Sn formation in thin films. By a similar argument the rate limiting factor of the Cu_6Sn_5 growth is the reaction of copper and tin.

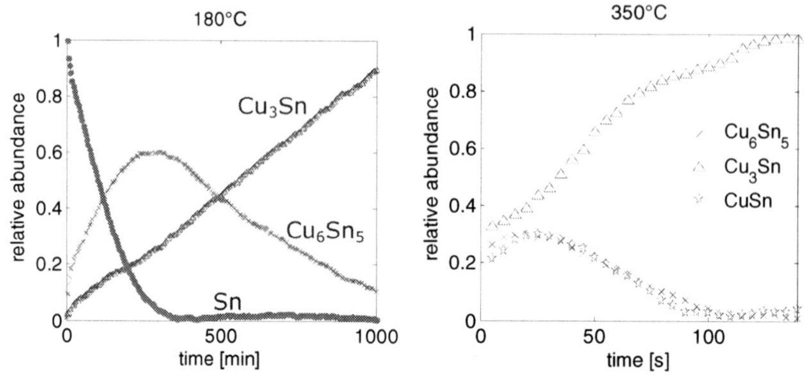

Figure 2: Evolution of the normalized, integrated intensities of the XRD peaks corresponding to Sn, Cu_6Sn_5 and Cu_3Sn at annealing temperatures of 180°C and 350°C. The tin is reacting with the underlying copper forming intermetallic phases. After 4 hours at 180°C all the tin is consumed and the ε–phase has grown at the expense of the η–phase.

In Figure 3 the formation of the ε–phase and the consumption of pure tin are plotted for different temperatures. By linear fits a reaction rate constant was extracted from the data for each temperature. According to the classical theory of the temperature dependence of reaction rate constants, the rate constant has an Arrhenius form,

$$k = Ae^{-E_a/k_BT}, \tag{1}$$

where k is the rate constant, A a pre-exponential factor, E_a the activation energy, k_b the Boltzmann constant, and T the temperature.

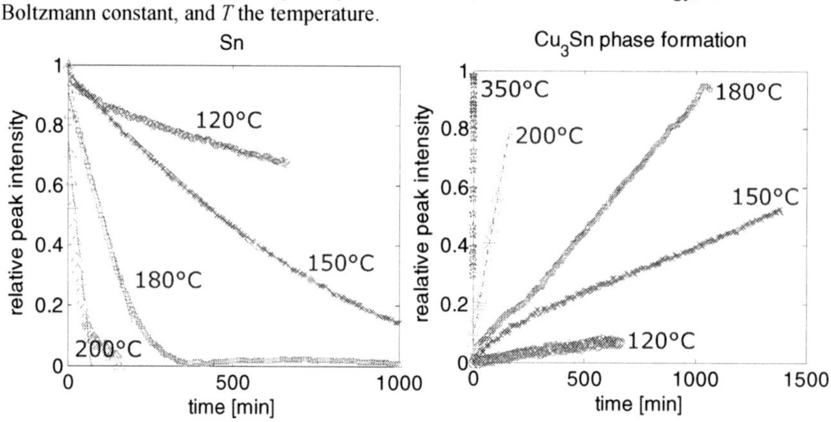

Figure 3: Consumption of Sn (left) and formation of Cu_3Sn (right) at different temperatures. The phase evolution is linear over a wide range.

A plot of $ln(k)$ versus $1/T$ yields a straight line with the activation energy E_a given by the slope (see Figure 4). The Arrhenius law holds very well even for temperatures above the melting point of tin. The activation energy for the formation of Cu_3Sn was determined to be 0.8 eV with a pre-exponential factor of $A=1.1\times10^3$ mol/s m, which is in good agreement with previous studies [9,14-17]. The activation energy for the consumption of tin, which is equivalent to the formation of Cu_6Sn_5, is slightly lower with 0.75 eV ($A==1.3\times10^3$ mol/s m). This also explains why at lower temperatures more Cu_6Sn_5 is formed than at higher temperatures.

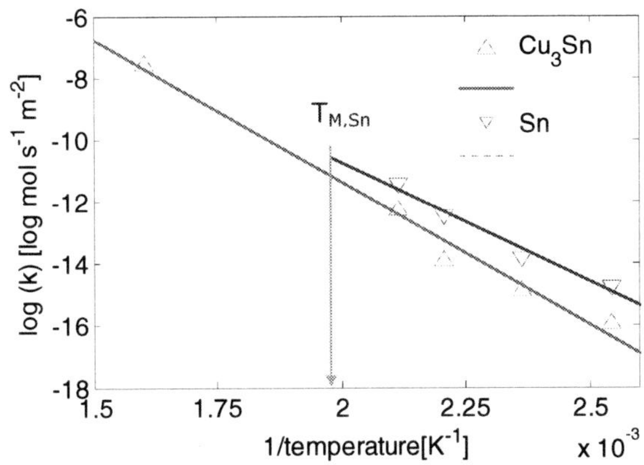

Figure 4: Arrhenius plot of logarithm of the reaction rate constants k versus $1/T$ for the consumption of tin and the formation of Cu_3Sn. The slope of the linear fit is equivalent to activation energy.

CONCLUSIONS

This study examines the kinetics of the reaction of Cu/Sn thin film couples as used in diffusion soldering. In-situ XRD was used to follow the formation of the Cu_3Sn and Cu_6Sn_5 intermetallic phases in real time. At 350°C a metastable CuSn phase was observed.

It was found that the rate limiting step for Cu/Sn diffusion soldering processes is not diffusion but rather the reaction kinetics at the Cu_3Sn/Cu_6Sn_5 interface.

The temperature dependence of the reaction rates could be described by an Arrhenius law with activation energies of 0.8 eV for Cu_3Sn formation and 0.75 eV for Cu_6Sn_5 formation.

REFERENCES

[1] D. Jacobson and S. Sangha, Soldering & Surface Mount Technology **8**, 12-15 (1996).
[2] D. Jacobson and G. Humpston, Soldering & Surface Mount Technology **4**, 27-32 (1992).
[3] K. Tu, Materials Chemistry and Physics **46**, 217-223.
[4] K. Tu, Acta Metallurgica **21**, 347-354 (1973).
[5] K.N. Tu and R.D. Thompson, Acta Metallurgica **30**, 947-952 (1982).
[6] S.K. Kang, R.S. Rai, and S. Purushothaman, Journal of Electronic Materials **25**, 1113–1120

(1996).
[7] S. Bader, W. Gust, and H. Hieber, Acta Metallurgica Et Materialia **43**, 329-337 (1995).
[8] A.Z. Miric and A. Grusd, Soldering & Surface Mount Technology **10**, 19–25 (1998).
[9] P.G. Harris and K.S. Chaggar, Soldering & Surface Mount Technology **10**, 38-52 (1998).
[10] P.A. Flinn and G.A. Waychunas, J. Vac. Sci. Technol. B **6**, 1749-1755 (1988).
[11] T. Massalski, (1996).
[12] A. Westgren and G. Phragmen, Z. Anorg. Chem **175**, 80 (1928).
[13] B.F. Dyson, J. Appl. Phys. **38**, 3408 (1967).
[14] M.E. Warwick and S.J. Muckett, Circuit World **9**, 5–11 (1993).
[15] P. Kay and C.A. McKay, Trans. Inst. Met. Fin. **57**, 169 (1979).
[16] P.J. Kay and C.A. McKay, Trans. Inst. Met. Fin. **54**, 68–74 (1976).
[17] C.J. Thwaites, Trans. Inst. Met. Fin. **43**, 143 (1965).

Mater. Res. Soc. Symp. Proc. Vol. 1318 © 2011 Materials Research Society
DOI: 10.1557/opl.2011.269

Stochastic Models of Epitaxial Growth

Dionisios Margetis[1], Paul N. Patrone[2], and T. L. Einstein[2]

[1]Department of Mathematics, and Institute for Physical Science and Technology, and Center for Scientific Computation and Mathematical Modeling, University of Maryland, College Park, MD 20742, U.S.A.

[2]Department of Physics, University of Maryland, College Park, MD 20742, U.S.A.

ABSTRACT

We study theoretical aspects of step fluctuations on vicinal surfaces by adding conservative white noise to the Burton-Cabrera-Frank model in one spatial dimension. We consider material deposition from above, as well as entropic and elastic-dipole step repulsions. Two approaches are discussed: (i) the linearization of stochastic equations when fluctuations are small, which captures correlations; and (ii) a mean field approach, which leaves out correlations but captures nonlinearities. Comparisons to kinetic Monte-Carlo simulations are presented.

INTRODUCTION

The fluctuations of steps on surfaces of crystalline materials such as silicon have been the subject of active experimental interest. A goal is to determine dominant pathways of atomic mass transport by observing the terrace width probability density or distribution (TWD) [1, 2]. Developing a complete theory of stochastic effects on vicinal surfaces poses a challenge. Crucial, yet largely unresolved, issues include: (1) the derivation of noise models from microscopic principles, and (2) the rigorous description of statistics for terrace widths in large systems. An approach that partly circumvents issue (1) is to add *ad hoc* noise to the celebrated Burton-Cabrera-Frank (BCF) model [3, 4] of step flow. In this vein, a systematic kinetic description of terrace-width fluctuations in terms of a mean field was proposed recently [5] on the basis of hierarchies for terrace correlation functions in one space dimension (1D). This formalism was recently applied to a large system of interacting steps in the absence of material deposition under the assumption that the noise obeys a second-order conservative scheme [6]. Here, we extend the formulation of [6] to the case with material deposition of flux F from above. In addition, we discuss possible physical implications that stem from the analysis, and illustrate limitations of approximations for the (intrinsically nonlinear) stochastic equations of motion. We show *how* the step interactions and deposition flux can jointly cause narrowing of the TWD, thus enriching the recent related work by Hamouda, Pimpinelli and Einstein [2] with step energetics.

THEORY: MODELING AND ANALYSIS

The step geometry is shown in figure 1. The j-th terrace is the region $x_j<x<x_{j+1}$ which has width w_j, where $j=0,\ldots, N-1$, and N is large. We apply screw periodic boundary conditions, mapping steps to point particles on a ring [5, 6]. For a vicinal crystal, the average terrace width is fixed at a constant value, $\langle w\rangle$, and w_j equals $\langle w\rangle$ initially (at t=0).

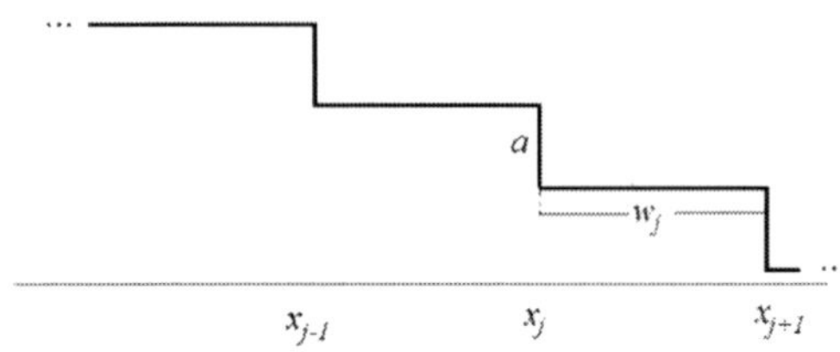

Figure 1. Schematic of a vicinal surface in 1D; x_j are step positions and a is the step height.

To formulate equations of motion for $w_j(t)$, we use the BCF theory [3, 4] with an external deposition flux, F, as well as entropic and elastic-dipole step repulsions [7]. Following [2], we apply a Galilean transformation to the co-moving frame with velocity $Fa\langle w\rangle$. We assume that step motion is slower than adatom diffusion (i.e., apply the quasistatic approximation) in this co-moving frame. Then, we add second-order conservative white noise to the equation of motion for w_j: the noise term is $-\eta_{j+1}+2\,\eta_j-\eta_{j-1}$ where $\eta_j(t)$ are independent (and identically distributed) white (delta-correlated) noises. This choice yields a *finite* variance of the TWD but is not unique. However, lower-order (non-conservative and first-order) schemes for white noise fail to yield a finite variance for long times [6]. Details of the formulation are given in [2, 5, 6, 8]. By change of variables from w_j to $w_j/\langle w\rangle$, denoted s_j, and use of non-dimensional time via scaling of t by some scale t^*, e.g., $\langle w\rangle^2/D_s$ where D_s is the terrace diffusivity, the stochastic equations are

$$\frac{ds_j}{dt} = -Fa\,\mathcal{A}(s_{j-2}, s_{j-1}, s_j, s_{j+1}, s_{j+2}) - \eta_{j+1} + 2\eta_j - \eta_{j-1}; \quad s_j = w_j\langle w\rangle^{-1}, \tag{1}$$

where $\mathcal{A}(\ldots,s,\ldots)$ blows up as $1/s^3$ if any terrace-width variable, s, approaches zero [6, 8].

Small fluctuations: linearization and steady-state TWD

The linearization of Eqs. (1) leads to the matrix equation

$$\frac{d\boldsymbol{s}}{dt} = -Fa\,\boldsymbol{A}\cdot\boldsymbol{s} + \boldsymbol{Q}\cdot\boldsymbol{\eta}; \qquad \boldsymbol{s} = \frac{\boldsymbol{w}}{\langle w\rangle}. \tag{2}$$

Here, $\boldsymbol{w}$ and $\boldsymbol{\eta}$ are N-dimensional vectors with components w_j and η_j, and $\boldsymbol{A}$ and $\boldsymbol{Q}$ are $N\times N$ circulant matrices. For example, for *diffusion-limited kinetics* these matrices have first rows [$1-2p+3g(1+2\varsigma)$, $-1+p-g(3+4\varsigma)$, $g(1+\varsigma)$, 0, ..., 0, $g\varsigma$, $p-g(1+4\varsigma)$] and [2, -1, 0, ..., 0, -1] where [8]

$$g = 3\frac{\tilde{g}}{k_B T}m_0{}^3(ac_s), \qquad \varsigma = \frac{1}{2}\frac{e^{-\nu}}{\sinh\nu}, \qquad p = \frac{1}{2}\left[\frac{\nu}{(\sinh\nu)^2} - \frac{e^{-\nu}}{\sinh\nu}\right], \qquad \nu = \frac{Fa\langle w\rangle^2}{2D_s}; \quad (3)$$

$\tilde{g}$ is a typical step-step interaction energy, k_BT is the Boltzmann energy, m_0 is the initial (constant) slope of the vicinal crystal, and c_s is the equilibrium adatom concentration at steps.

By basic stochastic calculus, we assert that the solution to Eq. (2) is a (vector) Gaussian random variable for every time $t>0$. Thus, the corresponding TWD turns out to be the Gaussian

$$P^{\mathrm{lin}}(s,t) = \frac{1}{\sqrt{2\pi\sigma_N(t)^2}}\exp\left[-\frac{(s-1)^2}{2\sigma_N(t)^2}\right]. \qquad (4)$$

For a finite number, N, of steps, the variance, $\sigma_N(t)^2$, of this TWD can be expressed exactly in terms of a discrete sum involving the eigenvalues of the matrices $\boldsymbol{QQ}^T$ and $\boldsymbol{A}+\boldsymbol{A}^T$ (where the superscript T denotes the matrix transpose) [5, 6]. Each set of eigenvalues is the discrete Fourier transform of the first row of the corresponding matrix [5]. In the macroscopic limit, $N\to\infty$, the discrete sum is replaced by an integral [5, 6]. The respective *steady-state* variance is found to be

$$\sigma_{\mathrm{st}}{}^2 = \sigma_\infty(t\to\infty)^2 = \frac{4}{Fa}\frac{1}{\sqrt{1-2p+4g(1+2\varsigma)}}\frac{1}{\sqrt{1-2p+4g(1+2\varsigma)}+\sqrt{1-2p}}, \qquad (5)$$

which accounts for terrace correlations (but not for nonlinearities). For zero deposition rate (i.e., as $F\to 0$), Eq. (5) yields $(2g)^{-1}\langle w\rangle^2/(D_s t^*)$, in agreement with [6]. In figure 2, we plot $\sigma_{\mathrm{st}}{}^2$ versus p for different values of the interaction parameter, g. Note that Eq. (4) predicts a nonzero probability for negative terrace widths, which amounts to step crossing. If $\sigma_{\mathrm{st}}{}^2$ is small enough, this likelihood of step crossing can be ignored. In [8], the variance of the TWD was computed for *flux-induced*, first-order conservative noise, whose coefficient depends linearly on p and stems from the asymmetric attachment of atoms at steps.

Mean field approach and decorrelation hypothesis

An alternate approach, in which nonlinearities are retained, is to fix the value of the neighboring terrace widths in each of Eqs. (1) to an *a priori unknown* mean field, $f(s,t)$. This field must be determined self-consistently [2, 5]. Seeds of this approach can be traced in the classic work by Gruber and Mullins [9]. Recently, the underlying kinetic framework was formulated systematically via appropriate hierarchies for terrace-terrace correlation functions [5].

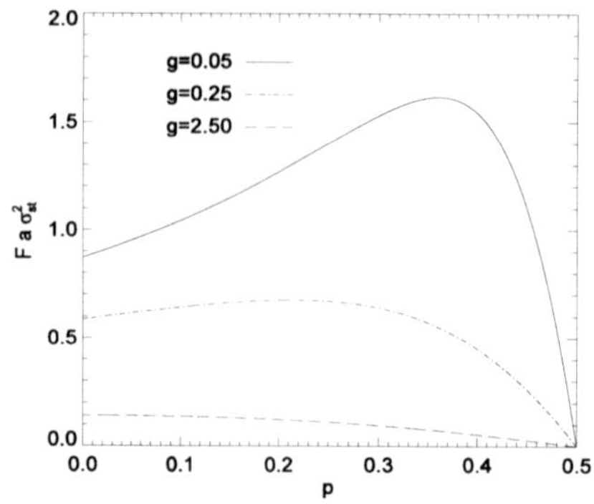

Figure 2. Plot of steady-state variance of the TWD as a function of parameter p by Eq. (5).

A goal is to reduce the system of Eqs. (1) to a single, tractable Langevin-type equation for an effective terrace width, $\hat{s}$ [5, 6]. This goal is achieved in three stages. *First*, the variables $s_{j\pm1}$ and $s_{j\pm2}$ are replaced by $f(\hat{s})$ and s_j is replaced by $\hat{s}$; and the noise term is replaced by $q\eta$, where q^2 equals 6 (the sum of squares of elements in the first row of matrix $\boldsymbol{Q}$) and η is the usual white noise. *Second*, we require that $\hat{s}$ produces the *same* TWD, $P(s)$, as the one from the starting system, Eqs. (1). Alas, the ensuing consistency formula for f turns out to be unwieldy: f is given in terms of the 5-terrace correlation function, $p^{(5)}$. *Third*, to enable analytical progress, we assume that the terrace widths are decorrelated, viz., $p^{(5)}$ is expressed as the product of the TWD's. By comparing the equations for the decorrelated s_j with the Langevin equation for $\hat{s}$, we derive an approximate formula for the self-consistent f. In the steady state, this formula reads [6]

$$\mathcal{A}\big(f(s), f(s), s, f(s), f(s)\big) \approx \iint_{-\infty}^{+\infty} \mathcal{A}(y_1, y_2, s, y_2, y_1) P(y_1) P(y_2)\, dy_1 dy_2. \tag{6}$$

For example, consider *diffusion-limited kinetics* without deposition [6]. For *large g*, Eq. (6) yields f approximately equal to unity (average s_j). The respective, "zeroth-order" TWD is

$$P_0^{\mathrm{mf}}(s) \propto s^{-\frac{4\breve{g}}{3}} \exp\left(-\frac{2\breve{g}s}{3} - \frac{\breve{g}}{3s^2} - \frac{4\breve{g}}{9s^3}\right), \quad \breve{g} = \frac{1}{3}\frac{t^* D_s}{\langle w\rangle^2}\, g, \quad s > 0; \quad P_0^{\mathrm{mf}}(s) \equiv 0, \quad s < 0. \tag{7}$$

By deriving an asymptotic expansion for f in powers of g^{-1}, we can obtain a more accurate, "composite expression" for the TWD that involves a single integral of $\mathcal{A}$ [6].

DISCUSSION

The analysis of stochastic effects in 1D reveals certain advantages of the mean field approximation over the small-fluctuation limit of the linearized model. Recall that the linearization of the governing equations of motion captures certain correlations (but cannot account for nonlinearities). In contrast, the mean field approach under the decorrelation ansatz captures nonlinearities but leaves out correlations. Thus, the question naturally arises as to which approximation is more reliable in the present setting of vicinal surfaces. To address this issue, we resort to comparing the TWD's predicted by the mean field and linearization approaches to the TWD produced via kinetic Monte Carlo simulations; see figure 3 [6]. This comparison indicates that the mean field approach, with a higher-order correction for f, reproduces accurately the essential features of our simulations for the steady-state TWD. A noteworthy feature is the *asymmetry* of the TWD, which should arise physically from the non-crossing condition for steps and is more pronounced as the step repulsion decreases. This asymmetry *cannot* be predicted by the linearized model, but is captured within the mean field approximation. Hence, in this vein, nonlinearities of step interactions tend to be more significant than terrace-terrace correlations. A similar observation can be made for the time-dependent TWD, although terrace correlations are stronger for finite times, and the above mean field approximation needs to be improved [6].

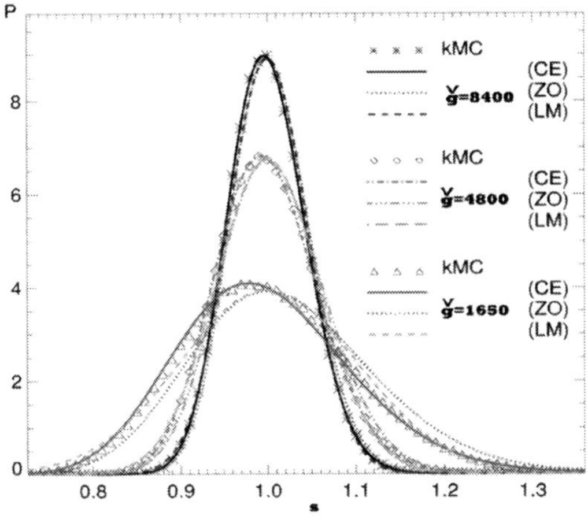

Figure 3. Steady-state TWD for F=0 and attachment-detachment limited kinetics, reproduced from figure 2 of [6]: (i) kinetic Monte Carlo simulations (kMC); (ii) linearized model (LM); and (iii) mean field approach by a zeroth-order (ZO) formula in which f is approximated by the average normalized terrace width ($f \approx 1$), and a composite expression (CE) in which f acquires a higher-order correction of the order of g^{-1}. The parameter $\check{g}$ is defined in Eq. (7).

Our analysis predicts the *narrowing* of the steady-state TWD under the combined effects of step repulsions and deposition flux, F. For example, consider diffusion-limited kinetics with recourse to Eqs. (4) and (5) as $N\rightarrow\infty$. For small F, the steady-state variance approaches a value that scales as g^{-1}. As F becomes large (i.e., $p\rightarrow0$), the steady-state variance decreases according to the asymptotic formula

$$\sigma_{\rm st}^2 \approx (Fa)^{-1}(1+4g)^{-\frac{1}{2}}\left(\sqrt{1+4g}+1\right)^{-1}. \qquad (8)$$

Results of the present study are limited in their applicability because of the assumed (one-dimensional) geometry. We expect that the detailed form of the TWD, especially for small values of the terrace width, can be distinctly different in two spatial dimensions (2D) where steps meander. For example, Eq. (7) reveals a singularity of the TWD at s=0. We believe that this type of singularity is an artifact of 1D. However, some of our results, e.g., the narrowing of the TWD with the step interaction strength, should hold for quasi-1D step systems.

CONCLUSIONS

This study explores the advantages, predictions and limitations of analytical approximation theories for terrace fluctuations on vicinal crystals in 1D. The mean field approach under a decorrelation hypothesis captures the asymmetry of the steady-state TWD, in contrast to the small-fluctuation limit of a linearized model. Both theories predict narrowing of the steady-state TWD with increasing deposition rate or step-repulsion strength. A limitation of the present study is due to the one-dimensional geometry. However, it is expected that some results, such as the scaling of the TWD with the deposition rate or the step-repulsion strength, should be applicable to quasi-1D systems.

ACKNOWLEDGMENTS

Research by the first author (DM) was supported by NSF under grant DMS 08-47587. Work by the other two authors (PP and TLE) was supported by the NSF MRSEC under grant DMR 05-20471 at the University of Maryland, with ancillary support from CNAM.

REFERENCES

1. S. D. Cohen, R. D. Schroll, T. L. Einstein, J. J. Métois, H. Gebremariam, H. L. Richards, and E. D. Williams, *Phys. Rev. B* **66**, 115310 (2002).
2. A. BH. Hamouda, A. Pimpinelli, and T. L. Einstein, *Europhys. Lett.* **88**, 26005 (2009).
3. W. K. Burton, N. Cabrera, and F. Frank, *Phil. Trans. Roy. Soc.* **243**, 299 (1951).
4. H. C. Jeong and E. D. Williams, *Surf. Sci. Rep.* **34**, 171 (1999).
5. D. Margetis, *J. Phys. A: Math. Theor.* **43**, 065003 (2010).
6. P. N. Patrone, T. L. Einstein, and D. Margetis, *Phys. Rev. E*, accepted for publication.
7. V. I. Marchenko and A. Ya. Parshin, *Sov. Phys. JETP* **52**, 129 (1980).
8. P. N. Patrone, R. Wang, and D. Margetis (unpublished).
9. E. E. Gruber and W. W. Mullins, *J. Phys. Chem. Solids* **28**, 875 (1967).

Mater. Res. Soc. Symp. Proc. Vol. 1318 © 2011 Materials Research Society
DOI: 10.1557/opl.2011.270

Characterisation of Organic Semiconductor Growth using Real-time Electron Spectroscopy.

D. Andrew Evans[1], Owain R. Roberts[1], Gruffudd T. Williams[1] and David. P. Langstaff[1]

[1]Institute of Mathematics and Physics,
Aberystwyth University,
Aberystwyth SY23 3BZ, U.K.

ABSTRACT

An organic molecular beam deposition system coupled to a soft x-ray excitation source has been developed to monitor the growth of organic semiconductor thin films in-situ and in real-time. Rapid collection of photoelectron spectra has been enabled using a multichannel array detector coupled to a hemispherical analyzer. The organic semiconductor tin phthalocyanine (SnPc) exhibits a Stranski-Krastanov growth mode on a polycrystalline gold substrate where the transition thickness between layered and clustered growth has been determined to be comparable to the thickness of a single molecular layer within which the molecules are standing on edge relative to the substrate plane.

INTRODUCTION

The electrical and optical properties of semiconducting organic materials in multilayer devices depend crucially on their physical and electronic structure and their interfaces with other materials within the device. For example, the conductivity of ordered small molecule films that have the highest mobility among organic semiconductors depends on the degree and direction of the molecular ordering within the film [1]. The molecular orientation and the larger-scale thin film morphology are sensitive to the substrate-molecule and molecule-molecule interaction at the interface and these are further influenced by thin film growth parameters such as temperature and rate. Metal phthalocyanines are widely applied in optoelectronic devices such OPV [2] and have other applications in fields as varied as cancer therapy and spintronics [3].

Few techniques can probe sufficiently broadly to provide direct correlations between these properties and the device performance and the usual approach is to apply different characterization methods sequentially on different device structures. One technique that can provide parallel information on energetics, chemical bonding and thin film morphology is photoelectron spectroscopy but it is not usually applied as a real-time, in-situ characterization method. However, with advances in electron detector and analyzer technology, photoelectron spectra can now be recorded sufficiently rapidly to enable data to be recorded during surface processing and thin film growth.

The growth of SnPc, a non-planar metallo-organic dye, on a gold surface is chosen to illustrate the application of in-situ, real-time photoelectron spectroscopy to the formation of an inorganic-organic interface.

EXPERIMENT

Exposure of a polycrystalline gold surface to a flux of the organic semiconductor SnPc has been monitored in real-time using photoelectron spectroscopy. The substrate surface was prepared by ex-situ polishing and solvent cleaning followed by in-situ Ar sputtering and annealing. Photoelectrons were excited by soft x-rays of energies ~ 100 eV and collected using a hemispherical analyzer coupled to a direct electron-counting array detector [4]. Excitation and collection were carried out during exposure to SnPc molecules, thermally evaporated from a calibrated and shuttered Knudsen cell, without changing the experimental geometry. Soft x-ray excitation ensured surface sensitivity and also enabled both band and core levels to be probed without changing the incident photon energy. This ensured an accurate energy scale throughout the measurement. A series of substrate and overlayer core levels (e.g. Au 4f, C 1s, N 1s and Sn 3d) as well as the high and low energy extrema of the electron distribution curve was collected sequentially by varying the energy of electrons entering the analyzer. The experimental arrangement is shown schematically in Figure 1. The electron detector contained 768 channels, each with its own amplifier and discriminator circuitry and was positioned behind a commercial multichannel plate at the focal plane of a hemispherical electron analyzer (Figure 1).

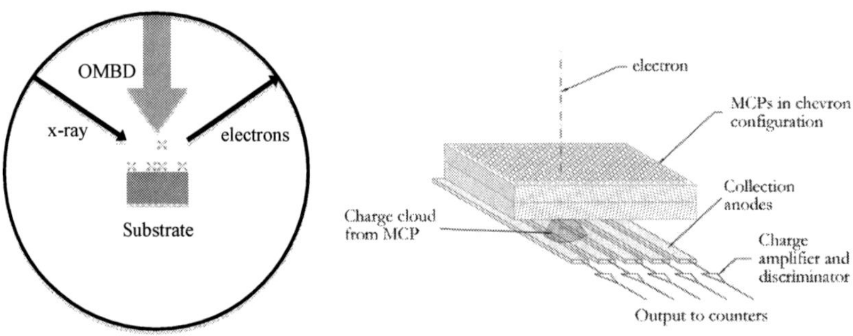

Figure 1: Real-time monitoring of organic thin film growth using photoelectron spectroscopy (left hand panel). Photoelectrons were collected using a multichannel array detector with an anode-anode pitch of 25 μm.

DISCUSSION

I. Temporal and energy resolution

A series of core level emission spectra are presented in Figure 2 to demonstrate the temporal and energy resolution of the technique. Panels A-C contain snapshot C 1s spectra for a diamond (001) surface terminated with oxygen using an acid etch. The composition and the structure of the surface were confirmed by conventional XPS and LEED [5]. A discernible peak was obtained in 25 ms (panel A) and it is possible to improve the data quality by averaging in the time domain as shown in panels B and C for a virtual integration time of 250 ms and 2.5 s respectively. These data collection times are adequate for many surface processing steps such as temperature cycling [5] and thin film deposition[6]. The dynamic range of the detectors ensures linearity for both low count rates (~ 10 Hz per pixel) to the maximum rate of 10^4 Hz per pixel. This enables low intensity and high intensity spectral features to be measured together and also enables large changes in peak intensity to be reliably monitored (for example during the growth of a thick overlayer). In panel D, a 4f doublet peak for the gold substrate is shown along with a conventional spectrum obtained by scanning the electron kinetic energy. The snapshot spectrum was recorded in 100 ms and reproduces all the essential features of the scanned spectrum that was recorded in 100 s.

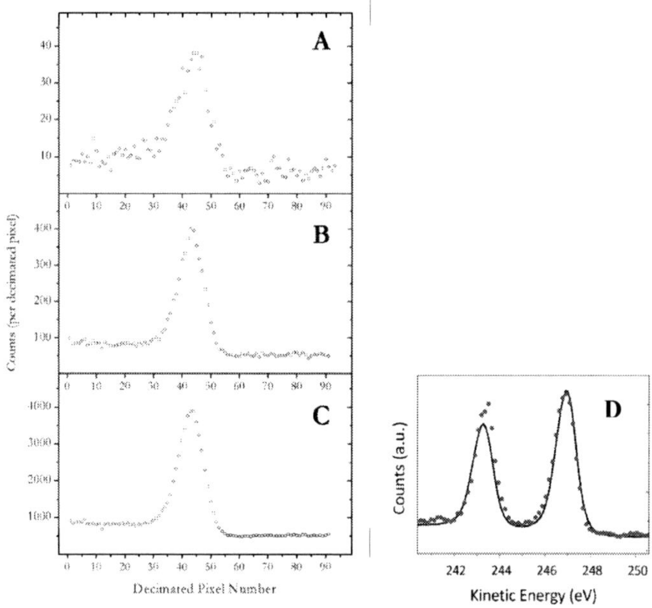

Figure 2: A 25 ms snapshot C 1s spectrum recorded for a clean diamond (001) surface irradiated with soft x-rays (A). The noise levels can be reduced by integrating in the time domain (B) and (C). A comparison of a Au 4f photoelectron emission spectrum recorded in snapshot mode across the array detector (circles) and in conventional scanning mode (D). The times taken were 100 ms and 100 s respectively.

II. Organic/inorganic interface formation

The adsorption of SnPc on a polycrystalline gold substrate can be followed by recording the time-evolution of substrate and overlayer core level peak intensities as shown for the Au 4f doublet in Figure 3 (left hand panel). In this 2-d projection of the data sequence, it is seen that there are no energy shifts or additional chemically-shifted components. Each peak was fitted to extract the intensity and the time (coverage) evolution of this intensity is also shown in Figure 3 (right hand panel). There is no change for the first 1000 s prior to the opening of the shutter of the SnPc cell. Due to the high surface sensitivity, the peak intensity is rapidly attenuated on exposure of the surface to the molecular flux, and the initial attenuation rate is exponential as expected for laminar growth. The slope of this curve provides a value of 1 nm for the mean free path of the Au 4f photelectrons that have a kinetic energy of around 250 eV.

There is a point of inflection at t = 1450 s that marks a reduction in the attenuation rate. This is due to an onset of molecular clustering at a temperature-dependent film thickness that has a value of 1.3 nm at room temperature. This value is close to the diameter of the SnPc molecule and this would suggest that the molecules are aligned perpendicular to the substrate within a layer of single molecules during the initial laminar growth stage. Beyond this layer, molecule-molecule interaction dominates over molecule-substrate interaction and the film morphology changes from layered to clustered growth.

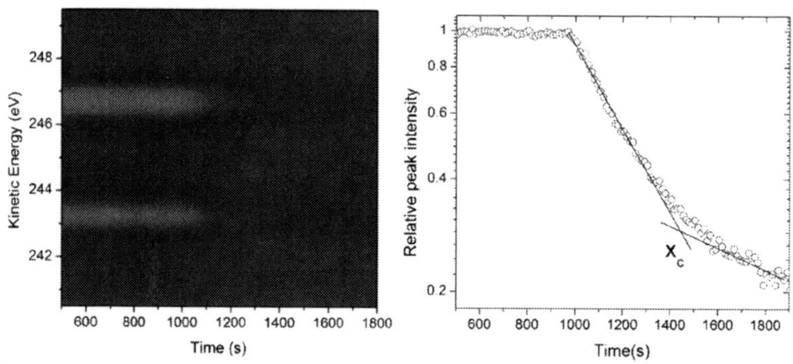

Figure 3: Time (coverage) evolution of the Au 4f core level for a clean polycrystalline gold surface exposed to a flux of the organic semiconductor SnPc (left hand panel). The intensity variation extracted from fitting the individual curves shows a distinct point of inflection at a critical thickness x_c (right hand panel).

CONCLUSION

Photoelectron spectroscopy has been developed as an in-situ, real-time probe of organic molecular beam deposition. The improved temporal resolution enables the growth morphology to be determined from peak intensity variation providing an accurate coverage for the transition from layered to clustered growth for SnPc on a polycrystalline gold substrate.

ACKNOWLEDGMENTS

These studies were carried out within the Centre for Advanced Functional Materials and Devices (CAFMaD), a HEFCW-funded Research and Enterprise Partnership. The work was also supported by the UK research councils (EPSRC and STFC).

REFERENCES

1. Steudel, S., et al., Nat Mater **4**, 597, (2005).
2. Li, N., et al., Appl Phys Lett **94**, 023307, (2009).
3. Dediu, V.A., et al., Nat Mater **8**, 707, (2009).
4. Langstaff, D.P., et al., Nucl Instrum Methods B **238**, 219, (2005).
5. Evans, D.A., et al., J Phys: Condens Matter **21** 364223, (2009).
6. Evans, D.A., et al., J Vac Sci Technol B **28,** C5F5, (2010).

Mater. Res. Soc. Symp. Proc. Vol. 1318 © 2011 Materials Research Society
DOI: 10.1557/opl.2011.271

The Electrical Conduction at Early Stages of Cluster-Assembled Films Growth

Emanuele Barborini[1*], Gabriele Corbelli[2] and Paolo Milani[2]

[1] Tethis SpA, via Russoli 3, 20143 Milano, Italy

[2] CIMAINA and Dipartimento di Fisica, Università di Milano, via Celoria 16, 20133 Milano, Italy

[*] emanuele.barborini@tethis-lab.com

ABSTRACT

Electrical conduction in Fe, Pd, Nb, W and Mo cluster-assembled films was investigated in-situ, during their growth by supersonic cluster beam deposition. We observed for cluster-assembled films resistivity values several orders of magnitude larger than corresponding bulk, as well as an increase of resistivity by increasing the film thickness, in contrast to the behaviour of atom-assembled metallic films. This suggests that nanoscale morphology arising by growth dynamics of cluster-assembled films, such as the minimal cluster-cluster interconnection and the evolution of surface roughness with thickness, may play a crucial role in the observed behaviour. Theoretical models based on non-isotropic 3D distributions of clusters into the film would help for a deeper understanding of the behaviour of cluster-assembled films compared to atom-assembled ones. Benefits are expected in the technological field of devices performing electrical read-out on active nanostructured layers, as in the case of chemoresistive sensors.

INTRODUCTION

A powerful approach for the synthesis of nanostructured films with tailored properties consists in the deposition of atomic clusters produced in gas phase. The survival of the nanoscale building blocks during the deposition process (memory effect), as well as film growth dynamics, can both deeply influence the transport properties of cluster-assembled systems [1-3]. In particular their electrical conductivity may differ significantly respect to conductivity of atom-assembled films, due to the presence of finite size effects, nanoscale disorder and peculiar topology evolution at early stages of film growth [4-7].

Although several studies on electrical conduction in cluster-assembled films at very low coverage [4-7], the role of cluster-surface interaction, fragmentation, diffusion, coalescence or sticking after the percolation onset and the first stage of 3D growth remains almost completely unexplored. A deeper knowledge in this field would provide benefits to many technological applications where the understanding of the influence of the nano- and mesostructure on transport properties are crucial. Examples are gas sensing through conductimetric microdevices, where the presence of chemical species in air is detected through variations of the electrical conduction of a nanostructured oxide film [8-12], as well as interconnections in highly integrated multi-layer electronics (3D electronics), where grain-boundary and surface scattering may affect electrical resistivity when the lateral dimensions are scaled down to nanometric length scales [13-16].

Cluster-assembled films have growth mechanism substantially different from atom-assembled ones so different relationship between their electrical conduction properties, morphology, and growth dynamics should be expected. To address this issue we studied in-situ the evolution of the electrical conduction of cluster-assembled thin films of Fe, Pd, Nb, W and Mo, during their growth by supersonic cluster beam deposition (SCBD).

EXPERIMENTAL DETAILS

Supersonic cluster beam deposition (SCBD) adopting a pulsed microplasma cluster source (PMCS) [2,17-18], was used for the production of cluster-assembled metallic thin films. Structure and operating principle of a PMCS were already described in [17]. Clusters are deposited at room temperature with a kinetic energy per atom of few tenth of eV, hence avoiding significant fragmentation of the aggregates. Regular deposition rates over a wide range of metals are routinely achieved by PMCS: in the present experiment we produced Fe, Pd, Nb, W and Mo.

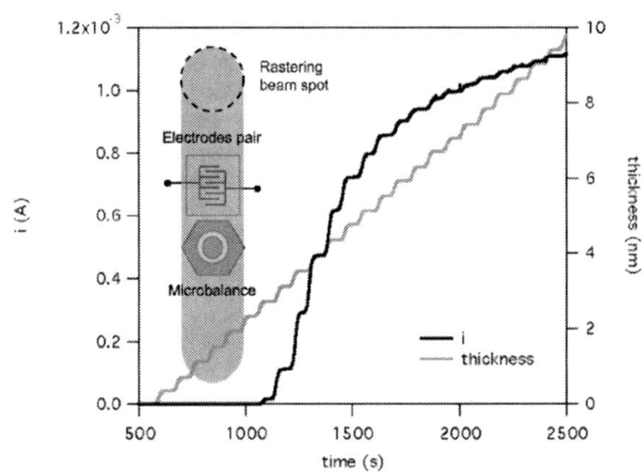

Figure 1. Data from electrodes and microbalance during rastered deposition. In the inset: scheme of the part of experimental setup for clusters collection, dedicated to the measurements of thickness and electrical conduction. In real experiment, beam spot and rastered area are much more larger than sketched.

Electrical conduction was measured in-situ during deposition on high resistivity alumina substrates with interdigitized electrode pair with 10 platinum fingers each (2 mm in length, 50 μm width and spacing) connected to an electrometer with internal voltage source (Keithley 6517A), applying a voltage of 0.1 V. Standard and high-grade surface polishing alumina were used to evaluate the role of substrate morphology. Alumina with electrodes was mounted close to a quartz microbalance on a motorized sample holder and rastered in front of the cluster beam spot, in order to simultaneously measure film thickness and conduction (figure 1 inset). In addition, substrate rastering allows the evaluation of the effect of background gas exposure (around 1×10^{-6} torr) on conduction, in time intervals where electrodes are not exposed to cluster beam. Figure 1 shows an example of raw conduction data and thickness data: upward steps in both current and thickness evolution are related to the periodic deposition of the cluster beam on the interdigitated electrodes and microbalance due to the rastering.

DISCUSSION

Figure 2a shows the behaviour of the electrical current across the cluster-assembled films, as a function of the thickness. Three separated regimes appear: i) no conduction below a certain critical thickness; ii) sharp onset of the conduction with current increasing of several order of magnitude occurring within a narrow thickness range; iii) asymptotic trend at larger thicknesses. The different regimes occur within almost the same thicknesses for all the metals. Figure 2b shows film resistivities as calculated from data of figure 2a. The graphs reveal resistivity values of 3-4 orders of magnitude higher than bulk ones for any metal at any thickness, as well as a non-monotonic behaviour, with a minimum around 5-6 nm. Remarkably the resistivity of cluster-assembled films has an increasing asymptotic trend, instead of a decreasing trend converging to bulk values, as observed in atom-assembled films [13,14,16]. Resistivity seems to increase according to a power law, as highlighted by the constant slope in the log-log plot, for all metals.

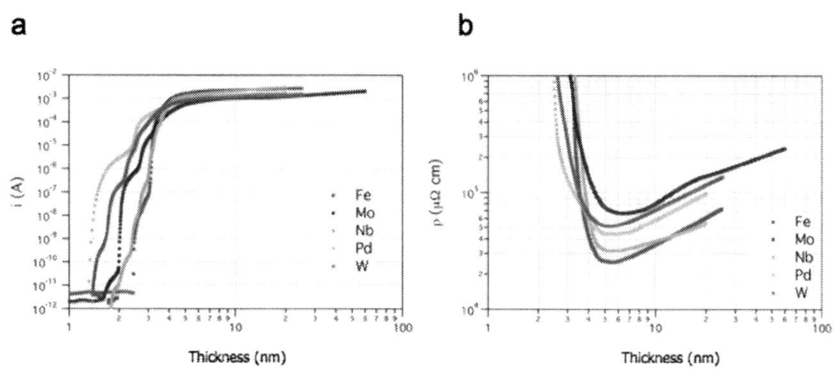

Figure 2. **(a)** Electrical current across the films as a function of the thickness. The log-log plot highlights the increasing of several order of magnitude of the current occurring in a few nanometers thickness range. **(b)** Resistivity of the cluster-assembled films as a function of film thickness. The non-monotonic behaviour is ascribable to peculiar topological evolution occurring in cluster-assembled film during the growth. The linear behaviour of asymptotic trend in the log-log plot indicates a power-law dependence.

The reaching of percolation threshold [19-21] may account for the sharp onset of the conduction. Once the first percolation path among deposited clusters completes, current starts flowing, and its subsequent huge increasing is due to the increasing of conduction paths becoming available as clusters progressively interconnect themselves during deposition. Within a small range of thicknesses, the distribution of the conduction phenomenon through the film evolves from 0D, to 1D, and to 2D. The 0D-1D transition takes place at a sub-monolayer coverage, the 1D-2D transition takes place between 1-2 and 3-4 nm. The current increasing slows down and resistivity reaches its minimum once 2D conduction saturates for the running out of in-plan conduction paths, at 5-6 nm. At this point, conduction within the film starts evolving in a 3D geometry.

CONCLUSIONS

We suppose that discrepancy of cluster-assembled film resistivity from that of atom-assembled films may be explained considering peculiar morphological elements and their hierarchical organization in the film. Minimal cluster-cluster interaction, due to low-energy conditions of deposition process, is expected to be at the base of percolative conduction. In addition, paths available for conduction should be shrunk by small contact areas, even when the 3D regime has been reached. All studied metals showed the same behaviour, suggesting that conduction features are largely dominated by morphology arguments, no matter the material in itself.

Peculiar 3D growth dynamics of cluster-assembled films ascribable to ballistic regime, whose main characteristic, beside minimal grain interconnections, is the progressive decreasing of topological connection of the uppermost layer [22] are supposed to explain the increasing asymptotic trend of resistivity. An evidence of this feature, highlighted by the increasing of surface roughness, has been reported in [23]. Therefore, even if the same amount of clusters is deposited during time, the contribution to electrical conduction gets lower and lower. We suppose that such "anisotropic" 3D percolation model, together with additional investigations at higher deposition temperatures (that should cause structure densification and increasing of cluster-cluster contact areas) could highlight these hypotheses and help to get a deeper knowledge in the electrical transport behaviour of cluster-assembled films with ballistic growth dynamics.

Beside fundamental insights on conduction process in nanostructured systems, the influence of nanoscale topology on the electrical resistivity may be of relevance, from the technological point of view, for the design of devices with electrical read-out where nanostructured layers are used as active elements, as in the case of chemoresistive sensors.

ACKNOWLEDGMENTS

We thank A. Podestà and M. Indrieri for AFM characterization, and S. Brown for stimulating discussion.

REFERENCES

1. A. Perez, P. Melinon, V. Dupuis, P. Jensen, B. Prevel, J. Tuaillon, L. Bardotti, C. Martet, M. Treilleux, M. Broyer, M. Pellarin, J.L. Vaille, B. Palpant, and J. Lerme, *J. Phys. D: Appl. Phys.* **30**, 709 (1997).
2. K. Wegner, P. Piseri, H. Vahedi Tafreshi, and P. Milani, *J. Phys. D: Appl. Phys.* **39**, R439 (2006).
3. P. Milani and S. Iannotta, Cluster Beam Synthesis of Nanostructured Materials, *Springer* Berlin (1999).
4. T. Hihara, Y. Yamada, M. Katoh, D.L. Peng, and K. Sumiyama, *J. Appl. Phys.* **94**, 7594 (2003).
5. S. Yamamuro, K. Sumiyama, T. Hihara, and K. Suzuki, *J. Phys.: Condens Matter* **11**, 3247 (1999).
6. P. Jensen, P. Melinon, A. Hoareau, J. Xiong Hu, B. Cabaud, M. Treilleux, E. Bernstein, and D. Guillot, *Physica A* **185**, 104 (1992).
7. J. Schmelzer, S.A. Brown, A. Wurl, M. Hyslop, and R. Blaikie, *Phys. Rev. Lett.* **88**, 226802 (2002).
8. M. Graf, A. Gurlo, N. Barsan, U. Weimar, and A. Hierlemann, *J. Nanoparticles Res.* **8**, 823 (2006).
9. E. Barborini, S. Vinati, M. Leccardi, P. Repetto, G. Bertolini, O. Rorato, L. Lorenzelli, M. Decarli, V. Guarnieri, C. Ducati, and P. Milani, *J. Micromech. Microeng.* **18**, 055015 (2008).
10. F. Favier, E.C. Walter, M.P. Zach, T. Benter, and R.M. Penner, *Science* **293**, 2227 (2001).
11. O. Dankert and A. Pundt, *Appl. Phys. Lett.* **81**, 1618 (2002).
12. T. Xu, M.P. Zach, Z.L. Xiao, D. Rosenmann, U. Welp, W.K. Kwok, and G.W. Crabtree, *Appl. Phys. Lett.* **86**, 203104 (2005).
13. W. Steinhögl, G. Schindler, G. Steinlesberger, M. Traving, and M. Engelhardt, *J. Appl. Phys.* **97**, 023706 (2005).
14. J.W. Lim and M. Isshiki, *J. Appl. Phys.* **99**, 094909 (2006).
15. C. Durkan and M.E. Welland, *Phys. Rev. B* **61**, 14215 (2000).
16. G. Steinlesberger, M. Engelhardt, G. Schindler, W. Steinhogl, A. von Glasow, K. Mosig, and E. Bertagnolli, *Microlectron. Eng.* **64**, 409 (2002).
17. E. Barborini, P. Piseri, and P. Milani, *J. Phys. D: Appl. Phys.* **32**, L105 (1999).
18. P. Milani, P. Piseri, E. Barborini, A. Podestà, and C. Lenardi, *J. Vac. Sci. Technol. A* **19**, 2025 (2001).
19. D. Stauffer and A. Aharony, *Thin Solid Films* **47**, 3 (1992).
20. N. Maheswar, G.S. Lodha, and R.V. Nandedkar, *J. Appl. Phys.* **100**, 113709 (2006) and references therein.
21. A.L. Barabasi and H.E. Stanley, Fractal Concepts in Surface Growth *Cambridge University Press*, Cambridge, England (1995).
22. A. Podestà, G. Bongiorno, P. Scopelliti, S. Bovio, P. Milani, C. Semprebon, and G. Mistura, *J. Phys. Chem. C* **113**, 18264 (2009).

Mater. Res. Soc. Symp. Proc. Vol. 1318 © 2011 Materials Research Society
DOI: 10.1557/opl.2011.625

Spreading kinetics at a molecular level

Jean-Luc Buraud, Olivier Noel, Dominique Ausserre
Université du Maine – LPEC –Molecular landscapes and biophotonics group, CNRS-UMR 6087, Le Mans – France

ABSTRACT

The spreading of a smectic nanodrop (8CB liquid crystal) on a solid surface was investigated by direct and real time imaging using the Surface Enhanced Ellipsometric Contrast (SEEC) microscopy [1-2]. The spreading ends with two molecular terraces (made of one monolayer and a bilayer). Two different behaviors were observed. In the first one the upper layer stays dense while shrinking. At the end of the process, the last molecules to disappear are located at the center of the initial disk. In the second one, nucleation and growth of holes is observed in the upper layer, in addition to shrinking. A model is proposed to describe the time evolution of the late stage structure. This model gives exact solutions of the kinetic equations, it covers strongly layered liquids such as smectic liquid crystals, it introduces the two dimensional Laplace pressure as an essential motor for spreading and it takes into account the liquid/gas transition in the surface layer that was consistently reported in experiments with 8CB. This model is in remarkable agreement with the experimental data and can explain the two observed behaviors [3].

INTRODUCTION

Finite extended molecular films on solid substrates present considerable interest both in technological and scientific research. On the fundamental side, they are particularly interesting as they provide finite 2D or quasi-2D thermodynamic systems[4-8]. They can be obtained by the deposition of a nanodrop in air, which evolves towards a molecular film by wetting a solid surface. In practice, the molecules used to make surface patterns are non symmetric and present, at least at the solid liquid interface, some amphiphilic character. The most extensively studied system is the combination of 4-n-octyl-4'-cyanobiphenyl (8CB) liquid crystal with oxidized silicon wafers. It may be considered as the model system of the field. Experimental studies on this system were mainly conducted using ellipsometry[9], Atomic Force Microscopy (AFM) [10] or X-ray reflectivity [11]. This behaviour was observed in the whole range extending from the solid/smectic transition temperature, 21.5 °C, up to about the nematic/isotrope transition temperature, 40.5 °C [12]. Even when the free surface of a macroscopic 8CB drop presents a finite contact angle with the solid surface, the molecular structure of the edge is made up of a sequence of flat terraces with a decreasing thickness in the outwards direction, lying on a surface precursor film [10].

During the spreading process terraces are disappearing one after each other from the top, while the lower terraces enlarge. The first visible terrace counted from the solid is a single bilayer over the monolayer terrace, the second one is made of several coupled bilayers, the next ones are again single bilayers.

EXPERIMENTAL AND MODEL

Experimental details

All the experiments reported below were conducted using the recent Surface Enhanced Ellipsometric Contrast (SEEC) technique [1, 2], which allows direct and real time visualization of molecular layers on surfaces, with a much higher resolution than ellipsometry or X-ray reflectivity. This technique commonly provides height resolution better than 0.1 nm, so that direct and real time visualization of molecular layers of 8CB becomes possible. Wetting experiments were conducted using 4-n-octyl-4'-cyanobiphenyl (8CB, 99.5 %, Merck) on oxidized silicon surfaces, for different temperatures ranging from 21.5 °C to 33.5°C. In this temperature range, 8CB is in a smectic phase. Optical observations were achieved through a polarization microscope (Leica DMRX PlanApo Pol Fluotar 20×0.5 objective lens). The solid support, a thin silicon wafer covered with a 106 nm thick oxide layer, was cleaned with UV/ozone and then, was set on a heating stage with ± 0.5 °C of accuracy. In order to install stable and reproducible experimental conditions, the sample was enclosed in a glove bag in which the relative humidity was kept constant at 30 ± 5% during the experiment. Pure 8CB was deposited onto the support using a home-made nano-spotter inserted into the glove bag. Microscope images were recorded through a tri CCD camera (Optronics DEI 750). Image processing was limited to shadow corrections.

In Figure 1a, we reproduce the image of the drop as it is seen directly from the microscope during the late stage. The bright patch is the bilayer, with radius R_2; the diffuse corona (radius $r > R_2$) is the monolayer in the gas phase. Figure 1b presents the schematic view of the internal drop structure that was described above. The different arrows in the figure represent the various material flows induced by spreading. Horizontal arrows demonstrate 2-dimensional intra-layer flows. Due to the radial drop shape, they have a radial symmetry. The vertical arrows demonstrate the permeation flux, a molecule per molecule material transfer from the bilayer towards the monolayer. The curved shape arrow figures how the molecules of the upper part of the bilayer, with zero radial velocity, pass into the lower part of the bilayer during bilayer shrinking. Figure 1c (the "Monument Valley butte") shows a topographic view of the drop right after the late stage structure was attained. The intensity values of Fig. 1a were converted into height values Z by inferring that image intensity is a regular function of the optical thickness of the liquid crystal [1]. This function was approximated by a second order polynomial that was determined using the values known from previous experiments [9, 13] i.e. 4.5 nm for the trilayer thickness, 1.2 nm for the monolayer thickness at the junction with the bilayer edge and zero thickness on the bare substrate.

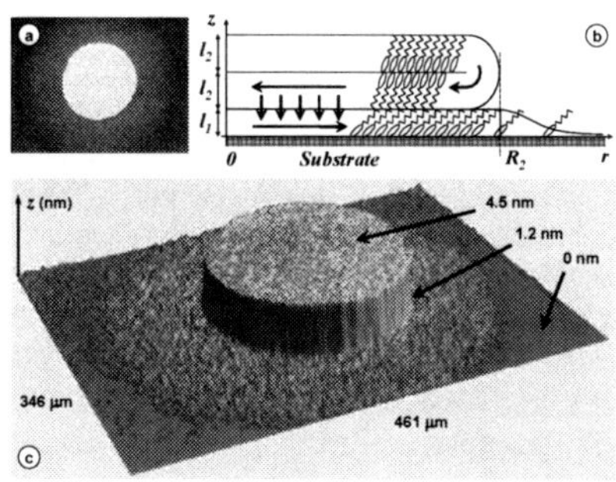

Figure 1. When the drop is very small, the entire drop structure is made of concentric terraces. a) optical microscopy image of a 8CB nanodrop in the late stage; b) scheme of the internal structure of the drop: a bilayer on the top of a monolayer; the monolayer is receding outwards; arrows represent the various material flows; c) 3D view of the optical thickness profile of the drop.

In Figure 2 we observe two series of top view images of 8CB bilayers that were arbitrarily extracted from kinetic experiments conducted at two different temperatures, respectively right above the solid/smectic transition temperature (22°C) and further away (26°C).

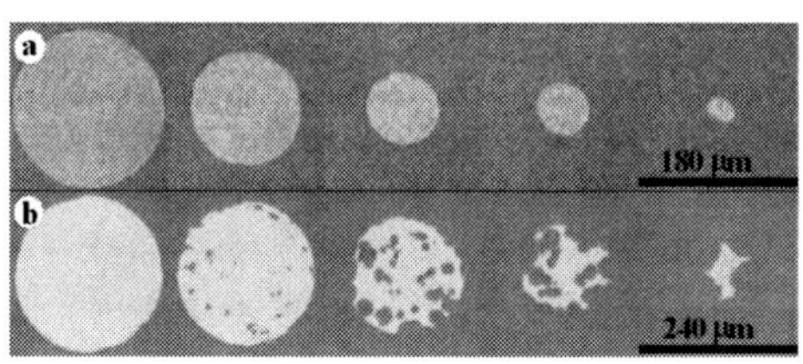

Figure 2. Bilayer evolution as observed a) upper sequence, at T=22 °C; b) lower sequence, at T=26 °C.

Both sequences start when spreading of the initial stratified droplet becomes reduced to the tri-layer structure. In the lower sequence b, nucleation and growth of holes is observed in the bilayer, in addition to shrinking [11]. Moreover, holes are not uniformly distributed over the layer but preferentially located in the outer part of the disk. Ostwald ripening and holes coalescence lead to the formation of one or several complex bilayer domains before completely vanishing. These patterns are conditioned by the random location of the holes that nucleate and therefore, are intrinsically unpredictable. In contrast, the deterministic evolution observed at low temperature allows direct comparison with theory. Then, we chose to focus on the low temperature regime as a priority.

Model

The spreading dynamics of stratified liquids was first described by de Gennes and Cazabat (dGC) [4]. It was originally intended to describe the behavior of simple liquids, which behave as stratified liquids at the contact of a solid surface, due to presmectic ordering induced by the surface. The dGC model describes the spreading dynamics of a terraced droplet on a wetting surface, which results in molecular steps. In this model spreading is driven by long range interactions between the liquid and the solid support. Each layer behaves as a 2D incompressible liquid, that is to say $\Sigma = Cte$ where Σ is the area per molecule. In this model we allows permeation currents between adjacent layers which is represented by interlayer current density $J_{n+1,n}$ $\left(m^3.m^{-2}.s^{-1}\right)$. Besides, the inner current density in the layer *n* is the horizontal velocity $\mathbf{V}_n$. There are also two more assumptions in the model, which are only implicit: i) Drops are large enough so that 2D Laplace pressure can be ignored; ii) The liquid is weakly stratified. The friction coefficients $\left(\zeta_{n-1,n}\right)$ between layers *n* and *n*-1 is supposed to be the same for every $n > 1$ value. It is given by

$$\zeta_{n-1,n} = \zeta = \eta \Sigma / l, \tag{1}$$

where η is the macroscopic viscosity of the liquid and l the thickness of the layer. Only the friction coefficient $\zeta_{0,1}$ between the first layer and the solid support takes a special value. This relationship is adapted for ordinary liquids. Equation 1 tells us that the friction between neighboring molecular layers is independent of the shearing direction. This requires that the diffusion coefficient is roughly isotropic, which is clearly not the case for smectics. A direct consequence of this relationship is that the permeation current between adjacent layers is allowed to be strong. Then, the essential part of the flow going out of terrace *n* with radius R_n takes place at the terrace edge in a permeation ribbon of radius *r* ranging between $R_n - \xi < r < R_n$, with a microscopic width $\xi << R_n$. Let us remember that permeation occurs by the hopping of individual molecules from a given layer to another one. It takes place across their common interface and it requires that these molecules overcome the energy barrier associated with the smectic potential. Here we claim that Equation (1) is no more valid for strongly stratified liquids such as smectic liquid crystals. Therefore the consequences of Equation (1) on the permeation current are not granted with these materials and actually, recent experiments conducted on smectic liquid crystals (8CB) in our group [3] do not conform the dGC model.

We propose a new model [14] which covers all systems and which is in particular compatible with the behavior observed on 8CB. First, this leads us to search a general solution to de Gennes – Cazabat equations which also covers strongly stratified materials. This solution takes into account the circular geometry of the drops. Secondly, the radius of a terrace disappearing during the kinetics goes down to zero. It is then necessary to consider the Laplace pressure due to the line tension of the domain edge as one among the driving forces of the dynamics. At least, we differentiate edge and surface permeation coefficients, because the molecules located in the defect line at a terrace edge must permeate more easily than the molecules located into the terrace.

We consider the case of structure made of one monolayer covered with one bi-layer. This case corresponds to most experimental situations encountered with smectic liquid crystals [11]. It is illustrated in Fig. 3, where one of the two possible structures of the dislocation line at the bilayer edge was arbitrary represented. This structure has no consequence on the results.
In Fig. 3 we have represented the net inter-layer or permeation current density $J_{21}\ (m^3.m^{-2}.s^{-1})$ volume from layer 2 towards layer 1, the horizontal velocity $\mathbf{V}_i$ and the thickness l_i in layer i.

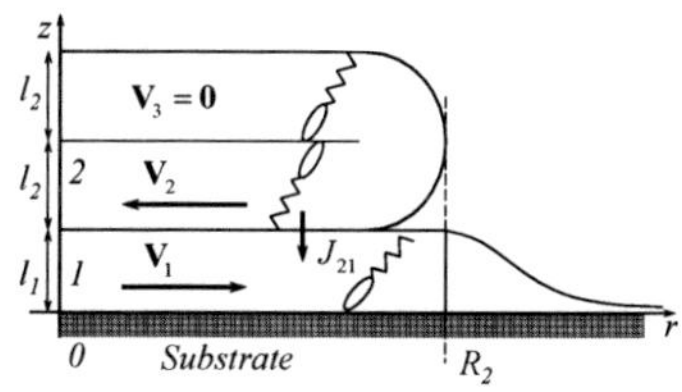

Figure 3. Bilayer structure: one bilayer on a dense monolayer with a liquid/gas transition at the bilayer edge. R_2 is the radius of the bilayer.

The comparison of the solutions of our model with experiments insure that $(\xi/R_2) >> 1$, where ξ is the permeation length. Hence, in our model, we get the kinetic law for bilayer shrinking:

$$\frac{dS_2}{dt} = -\alpha\left[\frac{(\tau_2-\tau_1)}{R_2}+\frac{W_2-W_1}{\Sigma}\right][S_2+\lambda P_2] \qquad (2)$$

Here S_2 and P_2 hold for area and perimeter of the bilayer. W_n is the attractive potential exerted by the solid on the liquid at level n and τ_n is the line tension at the edge of terrace n. Dimensional coefficients α and $\alpha\lambda$ are proportional to surface and edge permeation coefficients, respectively. Their ratio is characterized by the length λ. From equation 2 we can derive the time evolution of the bilayer radius R_2.

RESULTS AND DISCUSSION

Figure 4 shows on the same graph the experimental evolution of the bilayer radius and the evolution expected from our model. In the fit of figure 4 we have introduced three parameters: $T = \frac{2}{\alpha}\frac{\Sigma}{W_2-W_1}$, $L_1 = \frac{(\tau_2-\tau_1)\Sigma}{W_2-W_1}$, $L_2 = 2\lambda$. The best fit with our model was obtained for $L_1 = L_2 - 50\ \mu m$ and $T = 6840\ s$ [3].

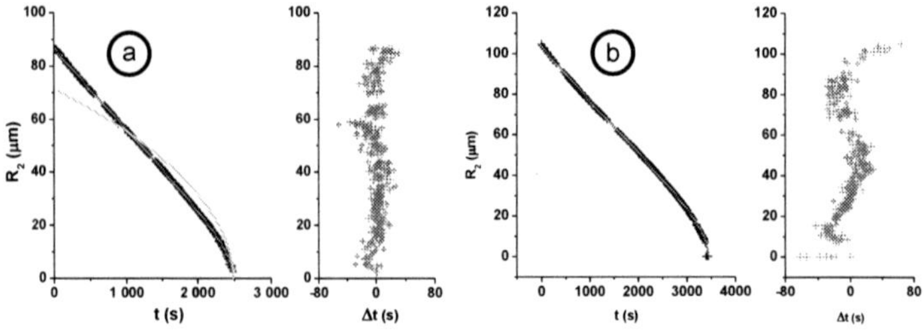

Figure 4. Radius of the last disappearing (shrinking) bilayer as a function of time. In a) (experiment reported in Ref. [3]) and b) (new experiment reported in the present paper), black cruces represent experimental data and the red line was obtained with the best fitting parameters in the new model. In a), the green line was obtained with the best fitting parameters in the dGC model. For each figure (a) and (b), the right side (red +) presents radius versus time distance Δt between experimental and calculated data.

Bilayer shrinking is driven by Laplace pressure and liquid solid interactions and involves two transport mechanisms, surface flow and line flow, with respective fluxes proportional to surface area and line length, the overall flow rate being regulated by the permeation coefficient. This remark makes clear that the limit $(\xi/R_2) >> 1$ corresponds to a quasi-static description of the kinetics. By contrast, the limit $(\xi/R_2) << 1$ would correspond to a stationary regime approach.

CONCLUSIONS

Although principally focused on 8CB liquid crystals, this work is expected to resume all the basic principles and useful equations for understanding the late stage spreading kinetics in stratified liquids. The main conclusion of our work is that one must recognize two asymptotic behaviors, corresponding respectively to weak stratification/high permeation regime (dGC model) and to strong segregation/weak permeation regime (8CB behavior).

ACKNOWLEDGMENTS

This work was supported by ANR under PNANO-07-050 project.

REFERENCES

[1] D. Ausserre, R. Abou-Khachfe, *Langmuir* 2007, 23, 8015-8020
[2] D. Ausserre, M. P. Valignat, *Nano-letters* 2006, 6, 1384-1388
[3] O. Noel, JL. Buraud, L. Berger, D. Ausserre *Langmuir* 2010, 26 (8), 6015–6018
[4] P. G. De Gennes, A. M. Cazabat, *C. R. Acad. Sci.* 1990, 310, 1601–1606

[5] P. Lazar, H. Schollmeyer, H. Riegler, *Phys. Rev. Lett.* 2005, 94, 116101-1–
116101-4
[6] W. H. De Jeu, B. I. Ostrovskii, A. N. Shalaginov, *Rev. Mod.* Phys. 2003, 75, 181 - 235.
[7] G. Oshanin, J. De-Coninck, A.-M. Cazabat, M. Moreau. *Phys. Rev. E* 1998, 58, R20-R23.
[8] D. T. Wasan, A. D. Nikolov, *Nature* 2003, 423, 156-159.
[9] S. Bardon, et al. *Phys. Rev. E* 1999, 596808-6818.
[10] L. Xu, Salmeron, M. & S. Bardon, *Phys. Rev. Lett.* 2000, 84, 1519-1522
[11] J. Daillant, G. Zalczer, J. Benattar, *Phys. Rev. A* 1992, 46, R6158-R6161.
[12] F. Vandenbrouck, *Thesis Universite-de-Paris VI*, **2001** Paris France, Paris.
[13] O. Benichou, et al. *Adv. Colloid Interface Sci.* 2003, 100-102, 381-398.
[14] D. Ausserre, JL. Buraud, *J. Chem. Phys.* To be published.

Novel Development and Applications of Scanning Probe Microscopy

Mater. Res. Soc. Symp. Proc. Vol. 1318 © 2011 Materials Research Society
DOI: 10.1557/opl.2011.423

Multiparameter Imaging and Understanding the Role of the Tip - Atomic Resolution Images of Rutile TiO_2 (110)

S. J. O'Brien, H. Ozgur Ozer*, G. L.W. Cross and J. B. Pethica
Centre for Research on Adaptive Nanostructures and Nanodevices (CRANN)
and School of Physics, Trinity College, Dublin 2, Ireland

ABSTRACT

A major challenge for scanned probe microscopy is to identify structures and chemical species on a surface, which have not already been inferred from other analytical techniques. Progress is impeded by the fact that in general the structure and composition of the tip atom is not known. To illustrate some of the issues involved, we report simultaneous scanning tunneling microscopy/atomic force microscopy (STM/AFM) of the TiO_2 (110) surface. The use of small amplitudes enabled the simultaneous acquisition of force gradient and barrier height images during standard STM imaging. Surprisingly, we find most STM images exhibit a corrugation contrast inverse to that usually reported in the literature. However, regardless of the contrast in STM, force gradient images always showed greater attraction over O rows. Barrier height images also show this consistency, always being greater over O rows. This supports the theoretical model of the electronic structure of the surface, but shows that the tip structure and interaction cannot be ignored in modeling STM images. We conclude that there is a fine balance between topography and local density of states (LDOS) in STM imaging of this surface; which of them dominates the STM image is determined by the tip. Simultaneous multi-parameter imaging is useful in interpreting images reliably, particularly on multi-component surfaces.

INTRODUCTION

Atom resolved scanning probe microscopy (SPM) of multi-element surfaces has the ability, in principle, to identify individual atomic species in their local chemical environment. However, its practicality is limited because the microscopic structure of the imaging tip, and its interaction with the surface in image formation, is in general not known. Elucidating the electronic and geometric structure of the imaging tip from experimental data can require extensive and laborious theoretical simulations using numerous tip models [1]. One possible way to improve matters is to perform both STM and AFM on the same surface. This gives information on different but still local properties, unlike the spatial averages obtained by other surface analytical techniques. The combination can constrain the range of possible tip and surface structures which fit any given image. Truly simultaneous operation of STM and AFM also eliminates ambiguities which arise from possible changes in tip and surface structure and in operating conditions between separate STM and AFM measurements.

We have previously shown that simultaneous STM/AFM is possible with sub-angstrom oscillation amplitudes [2]. The use of small amplitudes is important not only for a direct linear measurement of force gradients in non-contact AFM, but also for the proper acquisition of tunnel current and barrier height in STM mode. Measurement of the tunneling current in large amplitude AFM techniques is problematic, since the current is integrated over a distance much larger than the tunneling decay length [3]. Also, in general the peak current reached is much greater than in normal STM, which means strong interactions can dominate [4]. In this work we operate in normal STM mode, and acquire the force gradient and barrier as additional simultaneous parameters.

The TiO_2 (110) surface provides a good example of topography and electronic structure competing in determining STM image contrast. The (1×1) surface comprises alternate rows of Ti^{4+} and bridging O^{2-} atoms, running parallel to the [001] direction. Despite the topographic position of the bridging O atoms being higher then the Ti atoms, it is widely reported in the literature that when imaged under positive sample bias (i.e. imaging empty states) in STM, the Ti rows are imaged as bright. That is, the Ti atoms appear higher in STM images, due to the effect of LDOS. This consensus is based on numerous experimental reports ([5] and references therein) and theoretical calculations [6]. However, a few experimental reports of the inverse contrast can also be found in the literature [7–9]. These latter images show the bridging O rows as bright, or higher in STM images. Contrast inversion as a result of tip sample interactions in STM has been discussed in a number of theoretical studies [10–12]. The effect of tip electronic structure and tip-surface atom relaxations in STM has also been explored theoretically [13–15]. However, without any quantitative data on the forces acting on the tip during standard STM measurement there is usually not enough data to clearly define the tip. We now report simultaneous STM/AFM on the TiO_2 (110) surface, and show that there are wider variations of tip behaviour than are currently reported in the STM literature. Simultaneous AFM and barrier measurements cast some light on the nature of these variations and on the structure of the surface, independent of tip variations.

EXPERIMENTAL

A commercial UHV AFM/STM was modified by incorporating a Fabry-Perot fiber interferometer to give high sensitivity [16]. The cantilever was oscillated at ultra-small amplitudes (0.025-0.1 nm) at a fixed frequency well below resonance. Changes in the oscillation amplitude due to the force interaction were recorded using a lock-in amplifier. Using this amplitude data and the known stiffness of the cantilever, the force gradient is relatively straightforward to calculate [2]. Feedback was the same as standard STM, with the sample piezo displacement maintaining a constant tunneling current. Using a second lock-in amplifier the tunneling current oscillation dI due to the modulation dz was also recorded (see Figure 1). A value for the local apparent barrier height can therefore be calculated at every image pixel [17,18]. Both commercial Pt-Ir coated silicon tips (Nanosensors NCLPt), and homemade solid tungsten cantilevers and tips were used in the experiments reported here. The TiO_2 (110)-(1x1) surface was prepared by repeated cycles of Ar^+ ion sputtering and annealing. Typical STM parameters used during imaging were 0.1-0.2 nA tunnel current and 1.2-1.5V positive sample bias.

RESULTS

Two distinct STM imaging regimes were observed, easily distinguished by the

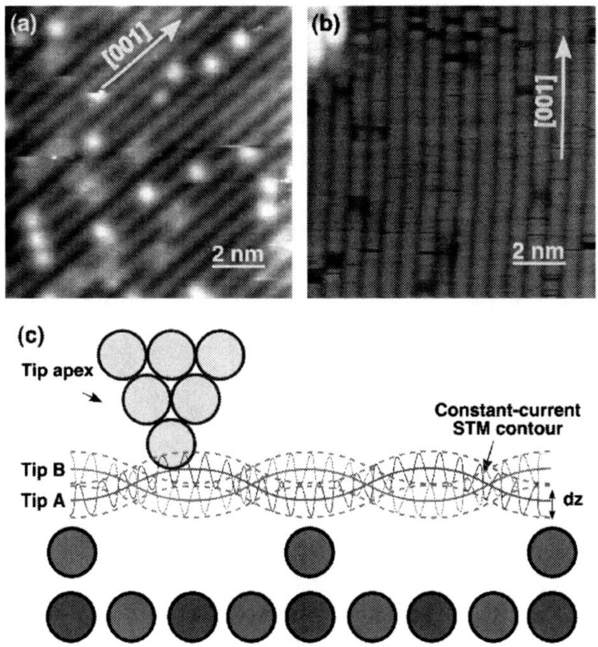

Figure 1. STM images showing imaging with tip type A (a) and tip type B (b). Panel (c) shows a simplified schematic (not to scale) of the measurement technique, and illustrates the different trajectories followed by the tip: type A which is dominated by LDOS and type B which follows the actual surface topography.

appearance of the bridging O row defects in STM. In the first of these, atomic scale defects appear mostly as bright objects between bright rows (Figure 1(a)). This type of contrast is widely reported in the literature and indicates that the Ti rows are imaged as bright. In the second regime, an inverted contrast is observed with the defects appearing as depressions on bright rows, indicating that the bridging O atom rows are imaged as bright (Figure 1(b)). Both regimes were observed with tungsten and with Pt-Ir coated silicon tips, and were stable for long periods of imaging, only changing after a tip crash. This strongly indicates that the different imaging contrasts are a result of distinct tip structures. We label the first type of image (Ti imaged bright) as due to tip type A and the second (O imaged bright) as due to tip type B. We have surveyed well over a hundred different images and find that the majority (~75%) of all STM images, using both Si and W tips, show the type B contrast. This ratio is unexpected, given that the current published literature on the subject shows mostly tip type A results. There are no previous literature reports on the relative incidence of tip types.

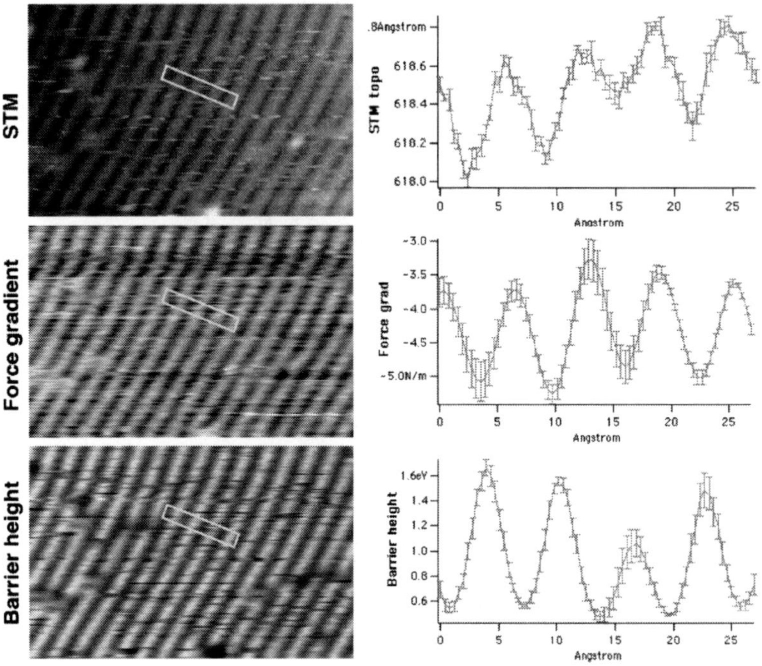

Figure 2. *(left)* Simultaneously acquired STM, force gradient and barrier-height data with tip type A. *(right)* Profiles averaged over the areas highlighted in the image.

Figure 2 shows an example of simultaneously acquired STM, force gradient and barrier height images taken with tip type A. The barrier image shows greater barrier height over O atoms compared to Ti atoms, in agreement with theoretical calculations and previous experimental results [19]. The force gradient image also shows bright defects between bright rows. Darker shading in the force gradient scan corresponds to larger oscillation amplitudes. The images therefore indicate a greater attractive force gradient over the O atoms. Figure 3 shows an example of simultaneous STM, force gradient and barrier height images taken with tip type B. The STM image is inverted compared with that shown in Figure 2, showing dark defects on bright rows. However, despite this STM inversion, the polarities of the force gradient and the barrier height contrast are unchanged compared to imaging with tip type A. This is most easily noticed in the contrast of the oxygen defects, which remain the same in the force and barrier images irrespective of tip type, but it is also obvious from the registry of rows and defects that the same applies to the main atom rows themselves. Evidently the underlying surface and defects are unchanged; the sign of both the force and barrier contrast is also unchanged, which suggests that the LDOS of the surface and its geometry is broadly unchanged. It is only the STM contrast that changes and inverts when the tip changes.

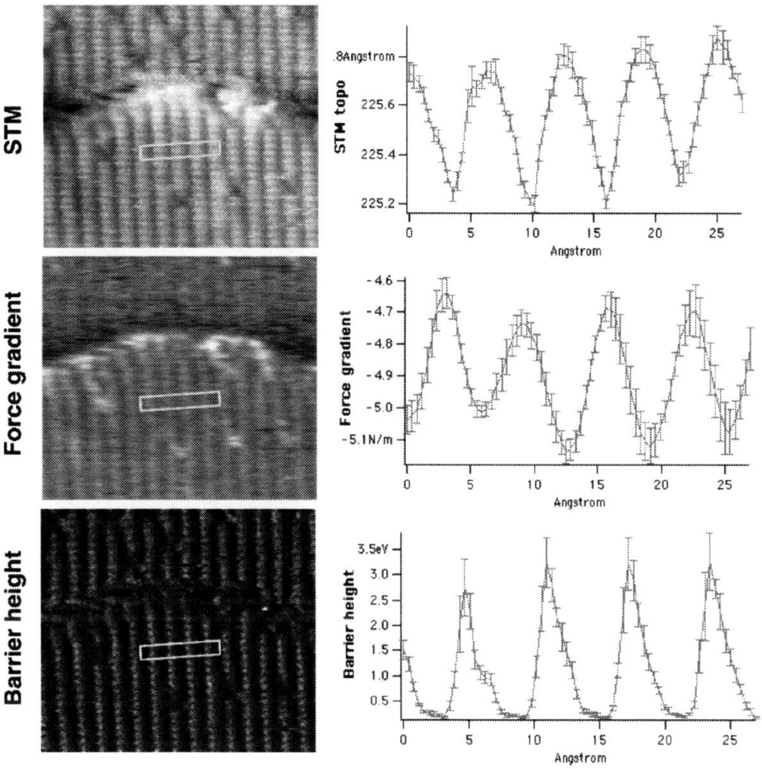

Figure 3. *(left)* Simultaneously acquired STM, force gradient and barrier-height data with tip type B. *(right)* Profiles averaged over the areas highlighted in the image.

For a set of 50 images, statistics were gathered on the measured corrugation values (i.e. site to site variation) in STM, force gradient and barrier height for both tip types A and B, and for W and Si base tip material. See Figure 4. The corrugation values for Si based tips fall in a much narrower range compared with the broad distribution of values for W tips. This suggests that for W tips there are a greater number of stable tip configurations that can give atomic resolution [20]. For tip type A the tip path is inverted with respect to the actual topography. Therefore, without tip atom relaxations, the corrugation contrast in force gradient and barrier height should be greater for tip type A than for tip type B, since corrugations generally decay with increasing distance from the surface. Averaging the corrugation values for Si tips shows this is indeed the case, implying that tip atom relaxations for Si levers are not significant components in the image contrast when changing from tip type A to B. For W tips, however, the trend is unclear since the change in mean corrugation is much less than the standard deviation of the average.

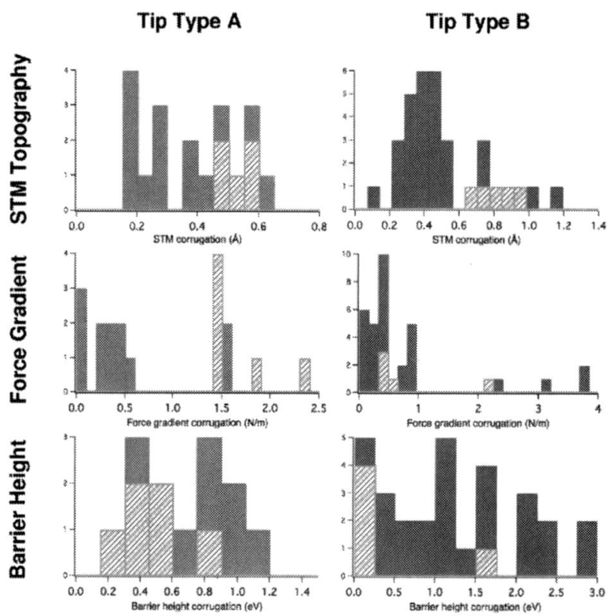

Figure 4. Set of histograms showing the range of STM, force gradient and barrier height corrugations observed for tip type A and type B. The distributions of corrugations for Si tips are shown hatched, distributions for W tips in solid.

DISCUSSION

These results raise a number of issues. Firstly, contrary to the current literature consensus on the subject ~75% of our STM images showed defects as dark objects between dark rows. This is the inverse of what is commonly presented and is not predicted by theoretical calculations [6] based on a Tersoff and Hamann (TH) model [21]. This suggests the presence of effects not considered by that model, such as tip electronic structure or significant tip-surface interaction. It has long been known that tip effects in STM require models beyond TH [22, 23]. Recent theoretical calculations which include tip interactions, have highlighted the sensitivity of STM contrast on the TiO_2 surface to tip changes [24]. However, despite the strong dependence of the STM contrast on the tip, our results show the contrast in both force gradient and barrier height are relatively stable, irrespective of tip type. The O^{2-} rows always appear more attractive in force gradient imaging, which indicates that the *total* charge density of the surface reflects the actual topography, as might be expected. Contrariwise, the LDOS corrugation near the Fermi level, which influences STM topography, does not reflect atom positions [6]; the decay rate must be faster over the O atoms. This is confirmed by barrier height imaging, where the O^{2-} rows appear as areas of greater apparent local barrier height, irrespective of tip type and in agreement with theory [6]. In short, whatever the tip structures are that cause inversion of the STM contrast, and hence of the actual path traced by the tip, they do not substantially affect contrast of either the force gradient or of the decay length of the surface charge density, and by implication the surface structure itself. This is encouraging, in that the tip interactions seem not to drastically modify the surface itself. Hence if the tip structure can be discovered, the images from the resulting modelling will be representative of the free surface.

For both tip materials the change in the tip apex from type A to B results in an increase in STM corrugation magnitudes of ~0.025 nm. In force gradient, however, the Si tips show a large drop in the measured corrugation values, whereas W tips show little change. This is a rather striking difference between the tip materials, which is not yet explained. The apex atom might well not be the same species as the tip bulk, but clearly the bulk tip material does affect either the species adsorbed or present as the local tip atom, and/or its interactions with the surface.

CONCLUSIONS

A novel multi-parameter scanning probe technique was applied to the TiO_2 (110) 1×1 surface. Inversions in the STM contrast were observed as a result of tip changes. However, the simultaneously imaged contrast in force gradient and barrier height did not invert, and also agrees with theoretical predictions and previous experimental evidence. This shows that the tip-surface interaction is not a minor perturbation in STM, and cannot be ignored in interpreting STM images. However, it does not seriously affect the force and barrier contrast, and by implication does not seriously perturb the actual surface structure. The results also demonstrate the advantages of having simultaneous multiple scanning parameters in diagnosing tip behaviour in SPM. The next step is to construct models of the tip interaction and resulting images, to ascertain if the combination of STM, force and barrier data enable a narrowing down of the possible tip structures, and perhaps therefore the definite determination of a unique structure. This would be an important step to enabling SPM to a-priori identify atomic species as well as geometries at surfaces.

ACKNOWLEDGEMENTS

This work was supported by Science Foundation Ireland Grant No. 00/PI.1/CO28 and the ESF EUROCORES FANAS program. H.O.Ozer would like to thank TUBA for travel support.

* Current address: Department of Physics, Istanbul Technical University, Turkey

REFERENCES

[1] Y. Sugimoto, P. Pou, M. Abe, P. Jelinek, R. Perez, S. Morita, and O. Custance, *Nature* **446**, 64, (2007).
[2] H. O. Ozer, S. J. O'Brien, and J. B. Pethica, *Appl. Phys. Lett.* **90**, 133110 (2007).
[3] C. L. Pang, A. Sasahara, H. Onishi, Q. Chen, and G. Thornton, *Phys. Rev.* B **74**, 073411 (2006).
[4] D. Sawada, Y. Sugimoto, K.-I. Morita, M. Abe, and S. Morita, *Appl. Phys. Lett.* **94**, 173117, (2009).
[5] U. Diebold, *Surf. Sci. Rep.* **48**, 53 (2003).
[6] U. Diebold, J. F. Anderson, K.-O. Ng, and D. Vanderbilt, *Phys. Rev. Lett.* **77,** 1322, (1996).
[7] S. Fischer, A. W. Munz, K.-D. Schierbaum, and W. Gopel, *Surf. Sci.* **337**, 17 (1995).
[8] C. J. Mitchell, A. Howard, M. Carney, and R. G. Egdell, *Surf. Sci.* **490**, 196 (2001).
[9] U. Diebold, J. Lehman, T. Mahmoud, M. Kuhn, G. Leonardelli, W. Hebenstreit, M. Schmid, and P. Varga, *Surf. Sci.* **411**, 137 (1998).
[10] G. Doyen, D. Drakova, and M. Scheffler, *Phys. Rev.* **B 47,** 9778 (1993).
[11] P. Jelınek, M. Svec, P. Pou, R. Perez, and V. Chab, *Phys. Rev. Lett.* **101**, 1, (2008).
[12] J. Nieminen, E. Niemi, and K. Rieder, *Surf. Sci.* **552**, 47 (2004).
[13] W. A. Hofer, A. J. Fisher, R. A. Wolkow, and P. Grutter, *Phys. Rev. Lett.* **87**, 236104 (2001).
[14] J. M. Blanco, C. Gonzalez, P. Jelinek, J. Ortega, F. Flores, and R. Perez, *Phys. Rev.* **B 70**, 085405 (2004).
[15] J. M. Blanco, C. Gonzalez, P. Jelınek, J. Ortega, F. Flores, R. Perez, M. Rose, M. Salmeron, J. Mendez, J. Wintterlin, and G. Ertl, *Phys. Rev.* **B 71**, 4 (2005).
[16] A. Oral, R. Grimble, H. O. Ozer, and J. B. Pethica, *Rev. Sci. Inst.* **74,** 3656 (2003).
[17] J. B. Pethica, J. Knall, and J. H. Wilson, *Inst. Phys. Conf. Ser.* **134,** 597 (1993).
[18] M. Herz, C. Schiller, F. J. Giessibl, and J. Mannhart, *Appl. Phys. Lett.* **86**, 153101 (2005).
[19] D. Ostermann, G. Walther, and K. D. Schierbaum, *Phys. Rev.* **B 71**, 235416 (2005).
[20] G. Cross, A. Schirmeisen, A. Stalder, P. Grutter, M. Tschudy, and U. Durig, *Phys. Rev. Lett.* **80**, 4685 (1998).
[21] J.Tersoff and D.Hamann, *Phys. Rev. Lett.,* **50**, 1998-2001, (1983)
[22] J.C. Dunphy, D.F. Ogletree, M.B. Salmeron, P. Sautet, M.-L. Bocquet and C.Joachim, *Ultramicroscopy* **42-44**, 490, (1992).

[23] J.H.Coombs and J.B.Pethica, *IBM J.Res & Dev.*, **30**, 455-459, (1986).
[24] G. H. Enevoldsen, H. P. Pinto, A. S. Foster, M. C. R. Jensen, A. Kuhnle, M. Reichling, W. A. Hofer, J. V. Lauritsen, and F. Besenbacher, *Phys. Rev.* **B 78**, 045416 (2008).

Mater. Res. Soc. Symp. Proc. Vol. 1318 © 2011 Materials Research Society
DOI: 10.1557/opl.2011.283

Experimental and Theoretical Study of the New Image Force Microscopy Principle

(Invited Paper)

H. Kumar Wickramasinghe and Indrajith Rajapaksa
Department of Electrical Engineering and Computer Science
University of California, Irvine, CA, 92697

ABSTRACT

A new technique in microscopy is demonstrated in which the domain of Atomic Force Microscopy (AFM) is extended to optical spectroscopy at the nanometer scale. Molecular resonance of feature sizes down to the single molecular level were detected and imaged purely by mechanical detection of the force gradient between the interaction of the optically driven object molecular dipole and its mirror image in a Platinum coated scanning probe tip. We provide full experimental details including a basic theory for this new technique. The microscopy and spectroscopy technique is extendable to frequencies ranging from radio to infrared and the ultra violet.

INTRODUCTION

AFM's have been successfully applied to nanometer scale imaging of chemical[1] magnetic[2,3] and electrostatic[4,5] properties of surfaces. These microscopes rely on probe tips (typically silicon) suitably modified to detect the specific property of interest and to translate it into a detectable force. AFM's have been highly successful for mapping surface topography – reaching atomic resolution. Several approaches have also been proposed for using AFM's to perform local optical spectroscopy of a surface. These include detecting light induced changes in contact potential difference[6] and detecting the photo thermal displacement due to absorption[7,8]. However these techniques have a major drawback in that they are all limited in resolution by the thermal diffusion length in the sample to around 0.25 micron. Recently, we introduced a new technique – Image Force Microscopy - where an AFM is modified to perform *molecular scale* spectroscopy and microscopy[9]. In this paper, we provide a complete description of the new microscopy principle. The ability to apply AFM's for nanometer scale spectroscopic analysis will open new opportunities in materials science and biology. In biology it could be used to study molecular cell surface interactions or for next generation DNA sequencing technologies for single base identification.

EXPERIMENT

Our microscope is built around an attractive tapping mode AFM[10] where the cantilever is vibrated at its first mechanical resonance f_0 (Fig. 1). We modulate the laser beam at a frequency f_m which modulates the image force gradient between object features and their mirror images also at f_m. This force gradient modulation in turn modulates the cantilever mechanical resonance frequency generating sidebands at f_0+f_m and f_0-f_m. The f_0+f_m sideband which is arranged to lie on top of the second mechanical resonance of the cantilever, is detected using a lock-in amplifier while the object is raster scanned to record the optical image. The reference for the lock-in is derived by mixing the f_0 and f_m drive signals in a double balanced mixer followed by a band pass filter centered at f_0+f_m. It should be pointed out that similar heterodyne approaches have been previously demonstrated in Kelvin Probe Force Microscopy[11] and Scanning Capacitance Force Microscopy[12].

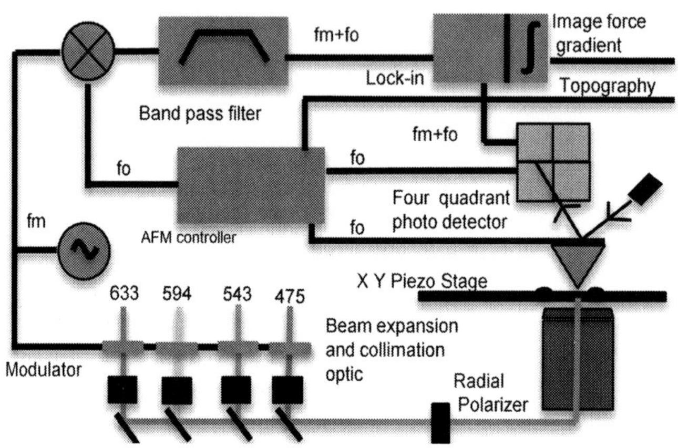

Figure 1. Schematic of the image force microscopy experiment.

A 10nm radius platinum coated silicon tip is illuminated by a laser beam using an oil immersion objective (NA =1.45) focused via a glass microscope cover slide. The focused laser excites an optical dipole moment within the object. As the tip approaches the object on the glass surface, this excited dipole sees its mirror image within the tip resulting in an attractive force and force gradient. To further enhance this dipole-dipole interaction, we use a radial polarizer (ARCoptix) at the entrance pupil of the objective. The optics is arranged so that we could independently couple up to four lasers (λ = 633nm, 594nm, 543nm, 475nm) into the objective. The optical power at the entrance pupil of the objective was adjusted to be 100μW. We used a cantilever with stiffness constant k = 3 n/m and first mechanical resonance f_0 = 65KHz. We chose a laser modulation frequency f_m = 360KHz and the frequency at the upper sideband f_0+f_m was detected at 425KHz.

In a series of experiments, with dye molecules on glass, we showed spectroscopy and microscopy at molecular resolution using this technique[9]. Here we show one such representative result where we imaged YOYO 3 dye islands (peak absorption near λ = 594nm) deposited on glass. We were fortunate in this experiment to trap a piece of debris within a YOYO 3 island. Fig. 2 (first row) shows a series of topography images recorded at 543nm, 594nm and 633nm

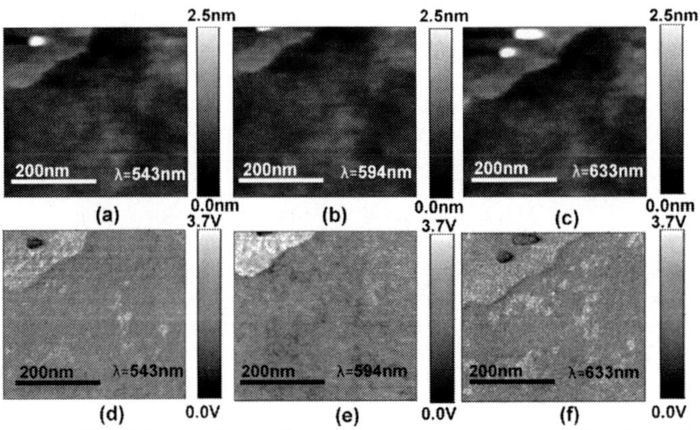

Figure 2. YOYO – 3 dye molecule images on glass as a function of wavelength: Topography images (a – c). Simultaneously recorded image force gradient images (d – f)

where the piece of debris in the center of the island is clearly visible. Fig 2 (second row) shows the simultaneously recorded optical image force gradient micrographs. The highest spectroscopic contrast is observed at λ = 594nm and as expected the debris particle shows no spectroscopic contrast.

Fig 3 illustrates a typical spectroscopy result. It shows the optical image force gradient signal on an island of 6-tamra dye that was deposited on a glass cover slide recorded as a function of wavelength superimposed on previously recorded Perkin-Elmer Spectrometer data (solid line); the agreement is excellent. The Perkin-Elmer data was recorded by spreading a very thick layer of 6- tamra dye on a glass microscope cover slide.

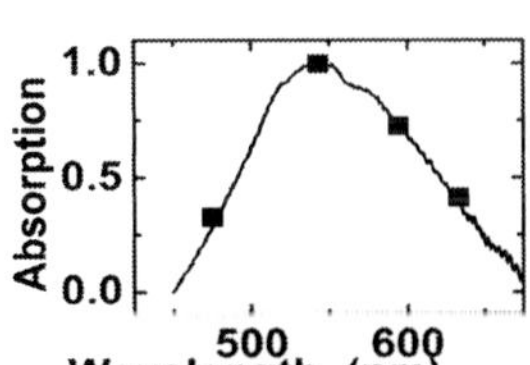

Figure 3. Spectral data points from image force microscope compared with bulk measurements from spectrometer

There are four possible candidates for the observed optical contrast. The first two candidates derive from photo displacement microscopy[13]. In the first case, the tip enhanced optical field can lead to a highly localized light absorption within the molecules beneath the tip at their resonance frequency resulting in a temperature increase and a corresponding periodic dilation of the glass substrate at the chopping frequency f_m. However, if this were the case, one would expect a dilation amplitude ΔL given by[14] $\Delta L = P_{abs}\alpha/\kappa$ where P_{abs} is the local absorbed power beneath the tip, α is the thermal expansion coefficient and κ is the thermal conductivity of the glass substrate. If we assume a single molecule under the tip with an absorption cross section $\sigma = 10^{-16}$ cm^2 we get $\Delta L = 6 \times 10^{-18}$ m – too small to be detected using our setup. In the second case, the the optical field can result in light absorption within the tip causing it to periodically dilate. We have both calculated and simulated this situation for our specific geometry of a platinum coated silicon tip with backside illumination and find that the resulting periodic dilation amplitude is less than 2.1×10^{-6} nm –too small to be detected by our system (details of this calculation are provided in the section on discussion and theory). The third candidate possibility results from the direct dilation of a molecule caused by the absorbed optical power. If we assume a rather high temperature increase of 10 K due to the molecular absorbed power and $\alpha = 10^{-6}$ /K we estimate the dilation of a molecule to be on the order of 2×10^{-6} nm – again too small to be detected in our setup. The fourth contrast mechanism is the one proposed in this article - namely the image force gradient between the object and its image dipole in the scanning probe tip.

To get an estimate of the image force gradients involved, we used the results of the finite element (COMSOL) analysis of the tip sample interaction. Details of this calculation can be found in the theory and discussion section. Here we present the salient features of the physics which can be understood by referring to Fig 4(a) showing a tip interacting with a particle on a substrate.

We can include the effect of the glass substrate by replacing it with the mirror image of the tip in the glass plus the mirror image of the particle in the glass – Fig 4(b); the complex reflectivity R of the glass mirror being given by the Fresnel equations $R = (1 - n)/(1+n)$, n being the complex refractive index of glass. We then approximate the situation in Fig 4(b) with Fig 4(c) where the particle/image pair in Fig 4(b) makes multiple reflections in the tip and in the (weak) tip mirror image. If we neglect the weak reflections, we can approximate the situation in Fig 4(c) to first order with Fig 4(d) where the particle is essentially interacting with its mirror image in the tip. Fig 5(e)- dotted line- shows the optical image force gradient versus tip-substrate spacing calculated using a full electromagnetic simulation of the situation in Fig 4(a) (without the particle) compared with experimental approach curves obtained with the control loop momentarily opened; the detected signal was normalized with respect to the measured tip vibration amplitude at the first harmonic. The vibration amplitude of the cantilever with the tip retracted far from the sample surface was 40 nm. Fig 4(f) shows the calculated force gradient versus tip – particle spacing for the case of Fig 4(a) with the particle in place. As expected, we see a rapid decay of the signal (over a distance of a nm) due to the highly localized image force dipole coupling. We also experimentally observe this rapid decay; the image contrast completely disappears when the tip is retracted just 1nm from the surface (much like in a tunneling microscope) although so far it has been difficult to record this decay curve due to the thermal drift in the scanning system.

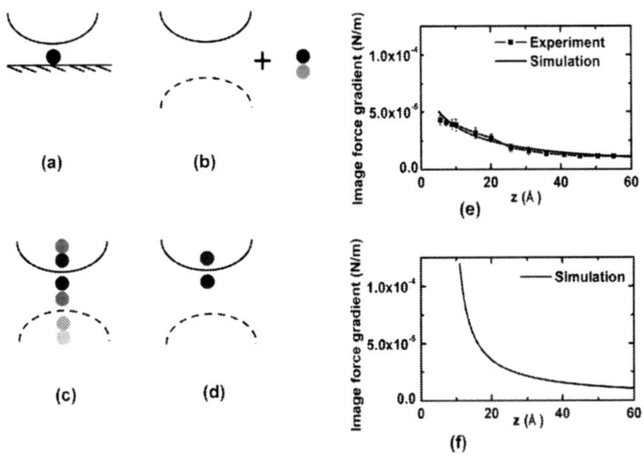

Figure 4. Experimental image force gradient approach curves compared with finite element (FE) simulations : Simplified representation of dipole-tip-sample interactions (a – d). Image force gradient on Pt tip approaching a glass substrate compared with FE simulations (solid line) -(e). FE simulation of image force gradient on Pt tip approaching a 1 nm particle on a glass substrate (f)

We can estimate the image force and force gradient from the measured data. The S/N ratio at the second resonance of 425 KHz was 30 dB in a bandwidth of 10 Hz. The Q at the second resonance Q_1 was 115 and the Q at the fundamental Q_2 was also 115. From this we calculate that the detected force was 2.7 x 10^{-12} N and the force gradient was 2.1 x10^{-5} N/m (see theory and discussion section). The detectable force can be reduced down to 10^{-18} N range by operating in vacuum and increasing Q to 10,000, by going to lower temperatures, by increasing cantilever resonance frequencies or by using lower stiffness cantilevers[10,15].

THEORY AND DISCUSSION

Electromagnetic field and image force simulation

We conducted extensive 3D simulations by solving Maxwell equations with the finite element package COMSOL .The incident wave (wavelength 543nm) was modeled as a plane wave with an electric field along the tip axis and an amplitude equal to that of the focused 100 μW laser spot.

Symmetry in the X coordinate was exploited to simplify the problem domain and increase memory efficiency. We modeled the Pt tip and glass substrate using published experimental material parameters[9]. The tip was modeled as an ellipsoid, height 0.29 μm and radius of curvature equal to that of the tips used in our experiment. The simulation domain was 0.5 x 1.0 x 1.8 μm. The infinite domain was simulated using 0.5 micrometers thick perfectly matched layers (PML). Care was taken to match the glass and air interfaces with the adjacent PML layers.

We considered two cases, the tip over the substrate and the tip over the particle on the substrate. The particle was a sphere of 1nm diameter containing 6 Tamra molecules. The absorption response of 6 Tamra was represented by an ideal Lorentzian oscillator at resonance[10]. From the Lorentzian model we extracted the complex dielectric constant 1.7 – 7.1i for the particle containing 6 Tamra molecules.

The z direction of the image force on the tip was calculated by performing a boundary integration of the Maxwell Stress Tensor over the tip surface [16].

$$F_z = \oint \sigma_{z,j} n_j \, ds \qquad (1)$$

Where the Maxwell Stress Tensor is given by

$$\sigma_{i,j} = \varepsilon_0 E_i E_j + 1/\mu_0 H_i H_j - \tfrac{1}{2} \delta_{i,j}(\boldsymbol{E}.\boldsymbol{E}+\boldsymbol{H}.\boldsymbol{H}) \qquad (2)$$

Finally, the force gradient was calculated by numerical differentiation of the Force vs Distance curve.

Analysis and estimation of tip heating effects

We use the COMSOL built in functions to calculate the time averaged (at optical frequency) deposition of power due to the electromagnetic absorption within the tip. Once again, we used parameters that matched our experiments. The tip consisted of a silicon core of radius 10 nm over coated with a 10 nm film of platinum. We first computed the electromagnetic field distribution as described in the previous section and then calculated the deposited power density within the tip. Fig 5 shows the color coded electric field distribution within the tip normalized with respect to the incident field.

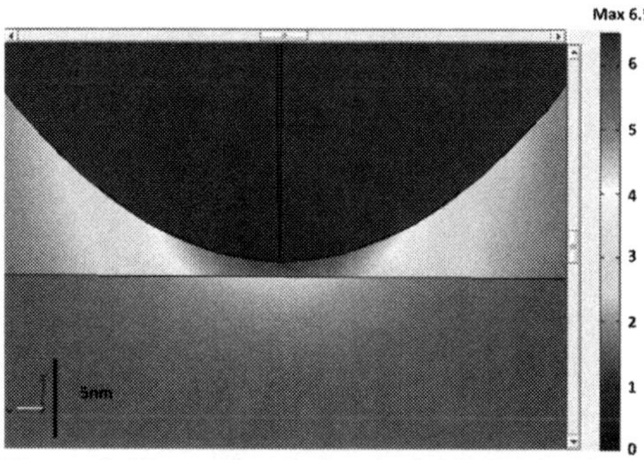

Figure 5. The scattered electric field(E)/ Incident field(E0) distribution of a Platinum coated Silicon tip 1nm above glass substrate

Fig 6 and 7 shows the color coded deposited power density distribution for the case shown in Fig 5. We note that the deposited power is highly localized to a hemispherical region of approximately 2 nm in radius at the very end of the tip. The average power density is approximately 1.0×10^{18} W/m^3 for the case where the tip is in contact with the glass sample (Figure 7) and 0.5×10^{18} W/m^3 when the tip is retracted 0.5 nm from the glass surface (Figure 8).We have validated our COMSOL simulations against the published work of Downes et. al[17] by repeating their simulations for the special case of a Gold tip on a Gold surface.

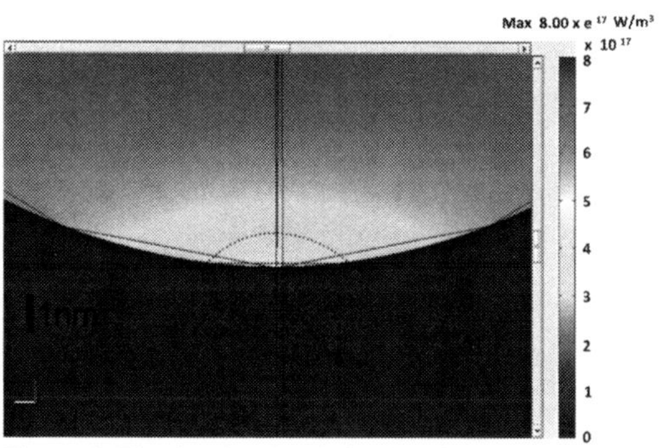

Figure 6 .The simulated resistive heating distribution in the Pt Tip touching the glass surface

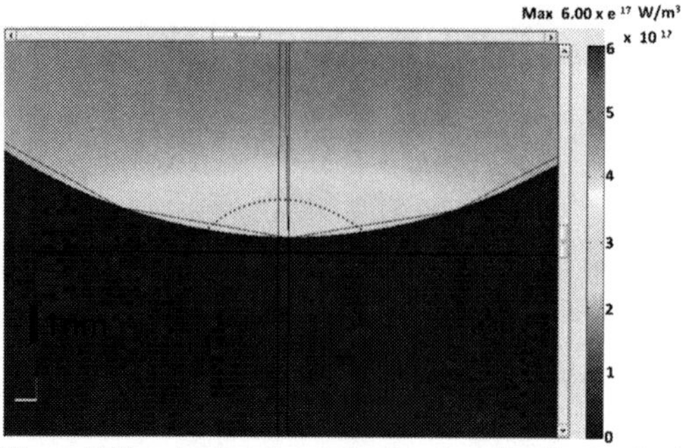

Figure 7.The simulated resistive heating distribution in the Pt Tip 0.5 nm retracted from the glass surface

Calculation of tip photothermal expansion

The deposited power is chopped at f_m (typically360 KHz in our experiments) in order to derive the Image Force Microscopy signal. This modulation results in a DC expansion plus an AC expansion of the tip at the chopping frequency. Since our contrast is derived from an AC signal, the constant DC expansion is of no consequence and will not be considered further. We follow the formulation given by Black et. al.[14] in the context of photothermal displacement spectroscopy to derive an expression for the AC component of the photothermal displacement. Fig 8 shows the tip geometry used in our derivation.

In this model, the heat is deposited at the very end of the tip and its modulation sets up a thermal wave that decays to 1/e of its initial value at a distance defined by the thermal diffusion length L_t given by

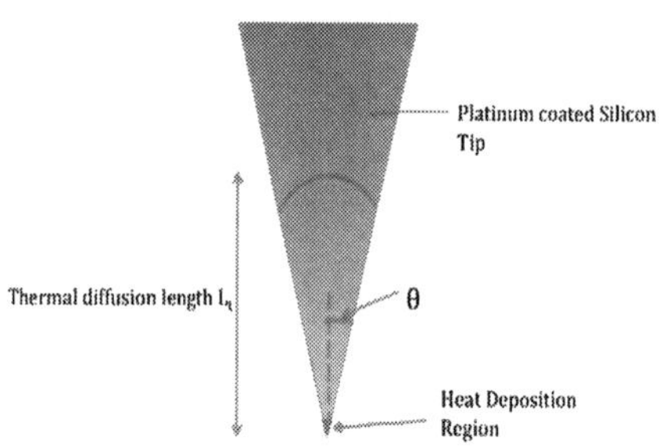

Figure 8. Tip geometry used in calculating photothermal displacement

$$L_t = \sqrt{[(\kappa/\rho C)(1/\pi f_m)]} \qquad 3$$

where κ is the thermal conductivity of the tip, ρ is the density, C is the specific heat and f_m is the optical chopping frequency. For the material properties used for our tips, L_t is

much larger than the size of the heat deposition region and therefore we are in the low frequency limit as defined in reference 5.
The temperature increase ΔT over a single cycle of modulation is given by

$$\Delta T = (P/2f_m)/ (\rho VC) \qquad 4$$

where P is the peak-peak modulation of absorbed power and V is the volume of the material being heated. In our case, the volume heated is a conical shaped region bounded by the tip end at the bottom and the thermal diffusion length at the top. The cone half angle is θ. V is given by

$$V = (1 - \cos\theta)\, 2\pi L_t^3 /3 \qquad 5$$

and the peak-peak temperature modulation is

$$\Delta T = \tfrac{3}{4} P / [\, \pi f_m \rho C(1 - \cos\theta)\, L_t^3] \qquad 6$$

This temperature modulation gives rise to a peak-peak tip expansion δL given by

$$\delta L = L_t\, \alpha\, \Delta T \qquad 7$$

where α is the thermal expansion coefficient of the tip. Substituting for ΔT yields the peak-peak expansion

$$\delta L = \tfrac{3}{4} P\alpha / [\, \kappa (1 - \cos\theta)\,] \qquad 8$$

We use the thermal parameters of Silicon rather than that of Platinum to estimate δL since the thermal diffusion length in Silicon is almost double that in Platinum. The absorbed power P deposited over the 1 nm radius hemisphere is $10^{-9} \times 2\pi/3$ W. If we take the thermal parameters for Silicon , $\kappa = 148$ W/mK , $\alpha = 4.7 \times 10^{-6}$ and a tip cone half angle of 10° , we get the peak-peak expansion $\delta L = 2.1 \times 10^{-6}$ nm
which is 100 times below the thermal noise of our cantilever.

We have also simulated the temperature increase and expansion of the tip by inputting the absorbed power from our finite element electromagnetic simulations into the COMSOL energy transport model for transient thermal analysis. The temperature distribution is shown in Fig 9. The calculated tip thermal expansion found by integrating the temperature increase over the 15 μm length of the tip is 4.1×10^{-7} nm - 5x smaller than the closed form expression value derived above.

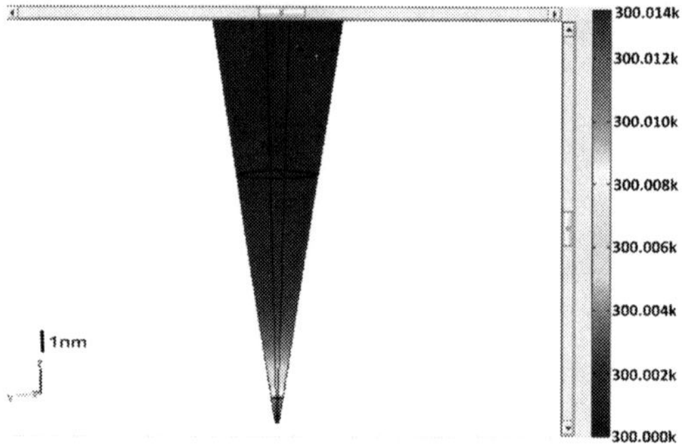

Figure. 9. Simulated transient temperature distribution in tip

Estimate of the detected force and force gradient

In this section we estimate the detected force and force gradient. The measured signal/noise at the second cantilever resonance of 425 KHz in a typical experiment is 30 dB in a bandwidth of 10 Hz. The thermal noise amplitude N of the cantilever at the second resonance is given by[10]

$$N^2 = 4KTBQ_2/\omega k_2 \qquad 9$$

where k_2 is the cantilever stiffness at the second resonance, K is Boltzmann constant, B is the system bandwidth, Q_2 is the Q of the cantilever second resonance, T is the absolute temperature and ω is the resonance frequency. For T = 300K, B = 10 Hz, Q_2 = 115, k_2= 128 N/m and $\omega = 2\pi$ x 425 KHz, we get N = 2.4 x 10^{-4} nm. The measured deflection X corresponding to the optical image force is 20 dB above this value or 2.4 x 10^{-3} nm and the corresponding force $F = k_2X/Q_2$ or 2.7 x 10^{-12} N. The smallest detectable optical force based on the cantilever noise N is 2.7 x 10^{-13} N.

We can reduce N and therefore significantly decrease the minimum detectable optical image force by going to lower temperatures, working in vacuum for higher Q's, using lower stiffness cantilevers or by going to higher cantilever frequencies.

We estimate the detected image force gradient as follows. The chopped optical beam at 360 KHz modulates the image force gradient between tip and sample which in turn modulates the cantilever oscillation amplitude at its first resonance or "carrier frequency" (65 KHz). This modulation results in "sidebands" about the carrier at 425 KHz and 295 KHz. If the modulation index is m, then the sideband to carrier ratio is m/2. The measured sideband to carrier ratio in our experiments was -80 db (after the Q enhancement at the second resonance has been factored out) so that m/2 = 10^{-4} and m = 2 x10^{-4}. The change ΔA in the cantilever oscillation amplitude A at its first resonance frequency caused by the optical image force gradient modulation F' acting on the cantilever is given by [10]

$$\Delta A/A = Q_1 F^{'} /2k_1 \qquad 10$$

where k_1 is the cantilever stiffness and Q_1 is the Q factor at the first resonance frequency. Since $\Delta A/A$ can be related to m by the relation $\Delta A/A = 2m$ the above equation yields $F^{'}$

$$F^{'} = 4m\, k_1 / Q_1 \qquad 11$$

CONCLUSION

We introduced a new technique of scanning microscopy based on detecting the image force gradient between an object and its mirror image in a scanning probe tip. While apertureless near-field microscopes[18,19] based on light scattering from tips have had great success in imaging nanoscopic objects from the visible to the infrared, the signals are very weak because they are based on far-field detection of a near-field interaction. We believe that our approach can be extended from the RF though the visible and infrared wavelengths and we see no impediment to achieving atomic resolution at visible wavelengths with this technique. Furthermore, electronic and magnetic oscillations could also be imaged using the same technique.

REFERENCES

1. C. D. Frisbie, L. F. Rozsnyai, A. Noy, M. S. Wrighton, and C. M. Lieber, Science **265,** 2071-2074 (1994).
2. Y. Martin and H. K. Wickramasinghe, Applied Physics Letters **50,** 1455-1457 (1987).
3. D. Rugar, H. J. Mamin, P. Guethner, S. E. Lambert, J. E. Stern, I. McFadyen, and T. Yogi, Journal of Applied Physics **68,** 1169 (1990).
4. Y. Martin, D. W. Abraham, and H. K. Wickramasinghe, Applied Physics Letters **52,** 1103 (1988).
5. M. Nonnenmacher, M. P. O'Boyle, and H. K. Wickramasinghe, Applied Physics Letters **58,** 2921 (1991).
6. M. Nonnenmacher and H. K. Wickramasinghe, Ultramicroscopy **42,** 351-354 (1992).
7. M. S. Anderson, Applied Spectroscopy **54,** 349-352 (2000).
8. A. Dazzi, R. Prazeres, F. Glotin, and J. M. Ortega, Infrared Physics & Technology **49,** 113-121 (2006).
9. I. Rajapaksa, K. Uenal, and H. K. Wickramasinghe, Applied Physics Letters **97,** 073121.
10. Y. Martin, C. C. Williams, and H. K. Wickramasinghe, Journal of Applied Physics **61,** 4723 (1987).
11. U. Zerweck, C. Loppacher, T. Otto, S. Grafström, and L. M. Eng, Physical Review B **71,** 125424 (2005).

12 K. Kobayashi, H. Yamada, and K. Matsushige, Applied Physics Letters **81,** 2629 (2002).

13 S. Ameri, E. A. Ash, V. Neuman, and C. R. Petts, Electronics Letters **17,** 337-338 (1981).

14 D. B. Eric, S. G. Ivan, R. R. Shanti, and G. L. Kenneth, Journal of Applied Physics **95,** 7655-7659 (2004).

15 D. Rugar, C. S. Yannoni, and J. A. Sidles, (1992).

16 J. D. Jackson, New York, Jhon Willey & Sons (1999).

17 A. Downes, D. Salter, and A. Elfick, spectroscopy **318,** 131-136 (2000).

18 F. Zenhausern, Y. Martin, and H. K. Wickramasinghe, Science **269,** 1083 (1995).

19 B. Knoll and F. Keilmann, Nature **399,** 134-137 (1999).

Mater. Res. Soc. Symp. Proc. Vol. 1318 © 2011 Materials Research Society
DOI: 10.1557/opl.2011.278

Crystallographic Processing of Scanning Tunneling Microscopy Images of Cobalt Phthalocyanines on Silver and Graphite

P. Moeck[1*], J. Straton[1], M. Toader[2], and M. Hietschold[2**]
[1] Nano-Crystallography Group, Department of Physics, Portland State University, Portland, OR 97207-0751, U.S.A, * pmoeck@pdx.edu, ** hietschold@physik.tu-chemnitz.de
[2] Institute of Physics, Chemnitz University of Technology, D-09126 Chemnitz, Germany

ABSTRACT

Monolayers of cobalt phthalocyanine (CoPc) and fluorinated cobalt phthalocyanine ($F_{16}CoPc$) on silver (111) and on highly (0001) oriented pyrolytic graphite (HOPG) were imaged with a scanning tunneling microscope (STM) at cryogenic temperatures (around 30 K) at Chemnitz University of Technology. Domains of regular arrays with periodicity in two dimensions (2D) and a variety of plane symmetries were observed. Crystallographic image processing (CIP) was used to quantify deviations from the plane symmetry groups and to obtain symmetrized versions of the content of the average unit cells of some of these arrays. Conclusions on the point symmetry of the CoPc and $F_{16}CoPc$ molecules within the arrays were drawn.

INTRODUCTION

Crystallographic image processing originated some 40 years ago with the electron crystallography community. Rather than solving crystal structures of unknowns on the basis of the weak phase object approximation of high resolution transmission electron microscopy [1], CIP has recently been used to "improve" images that were recorded in either the (atomic resolution) parallel illumination [2] or the Z-contrast scanning probe [3] modes of transmission electron microscopy. Such improvements concerned the visibility of individual atoms and of groups of atoms which were otherwise obscured by the imperfections (and signal to noise ratio) of the imaging process.

We present here the application of CIP to scanning tunneling microscopy images of 2D periodic array domains of (nominal) monolayers of CoPc and $F_{16}CoPc$ on Ag (111) and HOPG. As a result of reading this paper, other members of the scanning probe microscopy (SPM) community may become interested in using these techniques in their own surface science work. Since a comprehensive introduction of CIP for scanning probe microscopy (SPM) will soon be available in open access [4], our theory section will be very brief.

BRIEF DESCRIPTION of CIP for SCANNING PROBE MICROSCOPY

Having its roots in crystallography, the essence of CIP is averaging over many unit cells of a 2D periodic array in order to derive the content of one unit cell at a higher signal to noise ratio. For computational efficiency, this averaging is performed in Fourier space. The Fourier synthesis of the averaged Fourier coefficients of the SPM image intensity leads to the CIP processed (or symmetry enforced) SPM image in direct space. The well known Fourier filtering technique of SPM [5] can be considered the simplest form of CIP. That technique averages only over the translation symmetry in a 2D periodic array and is equivalent to the "CIP enforcing" of plane symmetry *p1*.

Sixteen of the 17 space symmetries that exist in 2D possess more symmetry elements than translations [6]. As a consequence of the multiplicity (which ranges from 2 to 12) of the so called general position, for the 16 higher symmetric plane groups CIP effectively averages over two to twelve times larger arrays than classical Fourier filtering. CIP also allows for the quantification of the deviation of a 2D periodic array from the mathematical definition of the 16 higher symmetric plane symmetries that are possible in nature. In order to quantify deviations from the 16 higher symmetric plane groups, the electron crystallography community uses Fourier coefficient amplitude (A_{res}) and phase (φ_{res}) residuals [1]:

$$A_{res} = \frac{\sum_{H,K} \left\| A_{obs}(H,K) \right| - \left| A_{sym}(H,K) \right\|}{\sum_{H,K} \left| A_{obs}(H,K) \right|} \quad \text{in \% and} \qquad \varphi_{res} = \frac{\sum_{H,K} w(H,K) \cdot \left| \varphi_{obs}(H,K) - \varphi_{sym}(H,K) \right|}{\sum_{H,K} w(H,K)} \quad \text{in degrees,}$$

where the subscripts stand for *obs*erved and *sym*metrized, w is a relative weight proportional to A_{obs}, and the sums are taken over all Fourier coefficients labeled H and K [1,4]. The amplitudes of Fourier coefficients that must be systematically absent when certain symmetry elements are present [6] form the basis of the so called Ao/Ae ratio [1]. For those plane groups for which it is defined [6], Ao/Ae ratios larger than zero quantify by how much an experimental SPM image of a 2D periodic array deviates from the respective groups.

The Fourier coefficient residuals, the Ao/Ae ratios, and subgroup/supergroup relations [6] weighted with geometric Akaike information criteria [7] form the basis for concluding which plane symmetry group an experimentally obtained SPM image from a 2D periodic array most likely possesses. The enforcing of this most likely symmetry is conceptually equivalent to the removal of the effects of the point spread function of the SPM from the experimental image. This function represents the combined effects of the "less then perfect" imaging process [4]. One may determine this function with the help of a highly symmetric calibration sample and then correct subsequently or previously recorded SPM images (that were recorded under essentially the same experimental conditions and with the same scanning probe tip) for the effects of this function. Also the signal to noise ratio of the SPM imaging process is enhanced due to effective averaging over two to twelve times larger arrays than in classical Fourier filtering. While work is in progress to quantify this enhancement, we expect that it will be proportional to the square root of the number of averages.

CIP ON STM IMAGES

F_{16}CoPc and CoPc on Ag (111)

The 3D point group of isolated CoPc and F_{16}CoPc molecules is *4/mmm* [4,8]. The orthogonal projection of this point group into 2D along the 4-fold rotation axis is *4mm*. Within a 2D periodic array, this is also the highest site symmetry that these molecules may posses, requiring that they be lying flat on the substrate and that they are not disturbed by interactions with the substrate and with neighboring molecules. The plane group *p4mm* possesses two positions with site symmetry *4mm* [6]. Subgroups of *4mm* are *4*, *2mm*, *2* and *1*. These point symmetries are site symmetries in the plane groups *p4*, *c2mm*, *p2mm*, *p2*, and *p1*. From both visual inspection and CIP, one can conclude that F_{16}CoPc and CoPc molecules are lying flat on strongly interacting Ag (111) substrates, Figs. 1 and 2. More precisely, both types of molecules arrange themselves on Ag (111) in plane groups *p2*, *p4*, and *p4mm*, possibly also in *c2mm*. The somewhat more unlikely arrangement in plane group *p2mm* would require a "two-molecule (periodic) motif" and relax the condition that all molecules in the array need to be identical. Only in plane group *p4mm* are the molecules likely to possess their full 2D point symmetry and the full degeneracy of electronic states that goes with this particular symmetry.

The (111) surface of Ag possesses the plane symmetry group *p6mm*. This plane group possesses the type I (translationsgleiche [6]) subgroup *c2mm*, which in turn possesses *p2* as one of its type I subgroups. Type I subgroups possess the same translations as their supergroup, but a lower number of site (i.e. point) symmetries within the unit cell [6]. For F_{16}CoPc and CoPc molecules that are lying flat on the substrate and are strongly interacting with it, a breaking of point symmetry elements of the isolated molecules is to be expected from Pierre Curie's well known symmetry principle [9]. Flat lying F_{16}CoPc and CoPc molecules may, therefore, arrange themselves on Ag (111) in plane groups *p2* and *c2mm*. Figure 1a shows two domains of F_{16}CoPc molecules with *p2* symmetry. Note that the domain in the lower part of the image possesses unit cell vectors that are nearly of the same length and that the angle between both vectors is neither close to 90° nor close to 60°. These are the conditions for centered unit cells in 2D. If the raw STM image were either corrected for the effects of the prevailing point spread function or an image with many more periodic motif repeats were available from this domain so that CIP processing would deliver meaningful results, it might have turned out that this domain actually possesses plane group *c2mm*.

Figure 1b reveals an arrangement of CoPc molecules in plane group *p4*. There are two positions with site symmetry *4* in this plane group. Only one of these positions is occupied by a CoPc molecule, resulting in a "one molecule motif". Unfortunately there are not enough periodic repeats of molecules in either of the two images, Figs. 1a and 1b, so that CIP could not be utilized to derive the averaged unit cell and to quantify the respective deviations from the plane symmetries. In agreement with Curie's symmetry principle [9], one can assume that the 2D point symmetries of F_{16}CoPc and CoPc in Figs. 1a and 1b are reduced to *2* and *4*, respectively.

Figure 2a shows sufficiently many repeats of the periodic motif so that CIP can be successfully applied. The results of *p4mm* enforcement on the basis of averaging over approximately 90 periodic motifs are shown in

Fig. 2b. Figure 2c shows for comparison the results of classical Fourier filtering. Table 1 gives an overview of the Fourier coefficient residuals for plane group *p4mm*, its type I subgroups and their respective type I subgroups. We use the (long) international Herman-Mauguin notation [6] in this paper. This allows us to specify the relative orientation of mirror and glide lines in plane groups *cm*, *pg*, and *pm*.

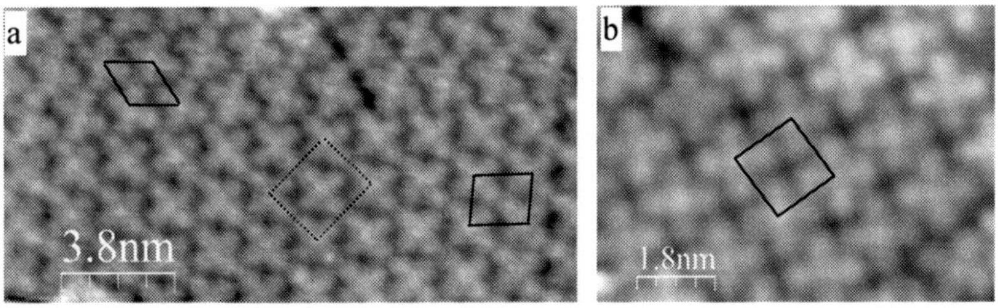

Figure 1. Raw STM images; **(a)** F_{16}CoPc on Ag (111), two *p2* domains with different oblique unit cells are shown. Primitive unit cells for these two domains are sketched in with full lines. An alternative centered unit cell is sketched in with dotted lines; **(b)** CoPc on Ag (111), one *p4* domain with one molecule per periodic motif is revealed. Imaging parameters: tunneling current $I_T = 80$ pA, tip bias $U_T = -1.5$ V for (a) and $I_T = 100$ pA, $U_T = -1.0$ V for (b). ($U_T < 0$ means tunneling from the sample into the tip of the STM.)

Figure 2. CoPc on Ag (111), $I_T = 150$ pA, $U_T = -1.0$ V; **(a)** Raw STM data with selection for the application of the CIP procedures; **(b)** *p4mm* enforced version of this data (on the basis of a circular selection of 512 pixels diameter that contained about 90 periodic motifs) with a quadratic unit cell inset as contour plot; and **(c)** translation averaged (*p1* enforced, i.e. Fourier filtered) version of the data (on the same basis) with an oblique unit cells inset as contour plot. The contour plot insets are rotated for convenience clockwise by approximately 8°.

Table 1: Fourier coefficient residuals for *p4mm*, its type I subgroups, and their type I subgroups.

	p4mm	***p4***	***c2mm***	***p2mm***	***c1lm***	***c1ml***	***p1lm***	***p1ml***	***p2***
A_{res} [%]	37.7	36.5	31.8	28.3	31.8	31.8	28.3	28.3	not defined
φ_{res} [°]	9.6	9.3	6.1	9.4	2.3	2.9	4.0	3.9	8.8

Note that the Ao/Ae ratio for both *cm* groups is zero. (The other groups in Table 1 do not possess such a ratio.) More than 50 Fourier coefficients with a dynamic (amplitude) range of 200 contributed to the residuals in Table 1. The average unit cell angle is 91.3°. The length of the unit cell vectors differ from each other by approximately 1 %. These deviations of the average unit cell from the definition of a quadratic unit cell are probably caused by the STM (e.g. due to thermal drift or creep of the piezoactuators) and are removed by the application of the CIP procedures when a quadratic lattice is enforced. It is quite reasonable to conclude from both Table 1 and the CIP averaged unit cell of Fig. 2b that the CoPc molecules arranged themselves in this particular case in a highly symmetric *p4mm* array. Since the molecules are located at positions of site symmetry *4mm*, they possess their full 2D point symmetry. In other words, the point symmetry breaking of flat lying F_{16}CoPc and CoPc molecules as inferred from Fig. 1 apparently did not happen in this particular case.

Note that there are two differently shaped "entities" with site symmetry *4mm* per unit cell in Fig. 2b. The larger one of these entities bears clear resemblance to CoPc molecules. The smaller entities should be CoPc molecules that are associated with a different set of adsorption sites that modifies the local density of states (LDOS) and, hence, their appearance in the STM image.

CoPc and F_{16}CoPc on HOPG

While individual flat lying CoPc molecules could be recognized on Ag (111) in Figs. 1b and 2 due to their four-fold lobes, this is not the case for the same kind of molecules on HOPG, Fig. 3a. From the viewpoint of Pierre Curie's symmetry principle [9], both adsorbate-substrate systems are equivalent. It is, however, well known that the CoPc molecules only interact weakly with a HOPG substrate so that point symmetry breaking and epitaxial relations do not need to occur when the molecules arrange themselves into a 2D periodic array.

Other authors inferred from a combination of "metastable-atom" electron spectroscopy and ultraviolet photoelectron spectroscopy measurements that copper-phthalocyanine (CuPc) molecules are not lying flat on HOPG at temperatures below 130 K [10]. This might be due to particularly weak substrate-adsorbate interactions at such temperatures.

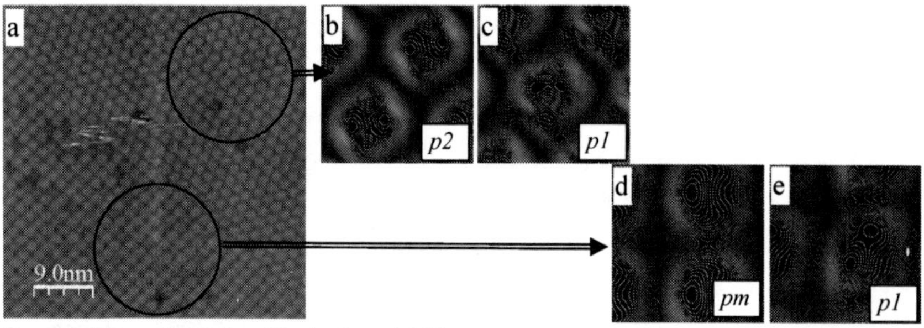

Figure 3. CoPc on HOPG, I_T = 80 pA, U_T = +1.0 V; **(a)** Raw STM data with two domains and selections for the application of the CIP procedures; **(b)** *p2* enforced motif as contour plot for the selected data from the upper domain (on the basis of the upper circular selection with a diameter of 126 pixels that contains about 85 periodic motifs); **(c)** translation averaged motif as contour plot for the same data (from the upper domain); **(d)** *pm* enforced motif as contour plot for the selected data from the lower domain (on the basis of the lower circular selection with a diameter of 126 pixels that contains about 90 periodic motifs); and **(e)** translation averaged motif as contour plot for the same data (from the lower domain).

Figure 3a shows that there are two different array domains. For the upper right part of this figure, CIP delivers an average unit cell with lattice vectors that differ by 8 % and are intersecting each other at an angle of 127.7°. It is reasonable to consider this lattice as being oblique so that the only possible higher symmetric plane group is *p2*, which possesses a phase residual of 15.9°. CIP was performed on the basis of approximately 85 molecules and 38 Fourier coefficients with a dynamic range of 200. Figures 3b and 3c show contour plots of both the *p2* and the translation averaged data.

For the lower left part of Fig. 3a, CIP delivers plane group *p1m1* with residuals A_{res} = 5.4 % and φ_{res} = 12.2° on the basis of averaging over approximately 90 molecules and 19 Fourier coefficients with a dynamic range of 200. (Amplitude residuals as low as this are often artifacts of low numbers of Fourier coefficients.)

The average lattice vectors of the respective unit cell differ by 3 % and intersect at an angle of 95.6°. Figures 3d and 3e show contour plots of both the *pm* and the translation averaged data. If the distortions to the image due to the prevalent point spread function of the STM were removed, this average lattice may be rectangular or even a square. The only site symmetry in plane group *p1m1* (besides *1*) is *m*, which is an oblique 2D projection point symmetry of isolated CoPc molecules. We speculate therefore that the molecules might perhaps not be lying flat in this domain. Note that the experimental results of other authors concerning CuPc on HOPG at cryogenic temperatures support this speculation [10]. For the upper domain with plane group *p2*, Fig. 3b, it seems reasonable to speculate as well that the molecules are not lying flat on HOPG. Note the differences of the results of CIP, Figs. 3b,d with respect to the results of classical Fourier filtering, Figs. 3c,e.

Figure 4a shows a similar analysis for $F_{16}CoPc$ molecules on HOPG. The raw STM image possesses 512 by 512 pixels and is sufficiently translation invariant that good results are to be expected from CIP. The (primitive) lattice vectors differ by 1 % in length and intersect at an angle of 116.3°. As the averaging was done over approximately 280 molecules (and 51 Fourier coefficients with a dynamic range of 200), these unit cell parameters seem to be sufficiently close to the definition of a hexagonal unit cell. The respective residuals for a central circular section of this STM image with a radius of 512 pixels are given in Tables 2 and 3.

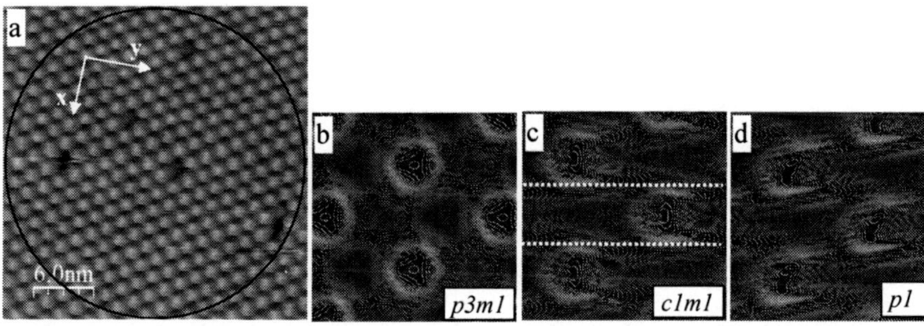

Figure 4. $F_{16}CoPc$ on HOPG, I_T = 150 pA, U_T = +1.0 V; **(a)** Raw STM data revealing only one domain; **(b)** *p3m1* enforced contour plot of approximately 4.5 unit cells (on the basis of the marked circular selection with a diameter of 512 pixels that contains about 280 periodic motifs); **(c)** *c1m1* enforced contour plot of approximately 2.25 (centered) unit cells, the dashed lines represent the glide lines of this group; and **(d)** translation averaged contour plot of approximately 4.5 unit cells. The CIP derived contour plots are all rotated for convenience by approximately 13° counterclockwise with respect to the coordinate system of the raw STM image (and are all based on the same circular selection of the experimental raw data).

Table 2 shows the residuals for *p6mm*, all of its hexagonal subgroups, and its higher symmetric oblique subgroup *p2*. (Ao/Ae ratios are not defined for all of these groups). Of the hexagonal subgroups, the residuals for *p3m1* are rather low considering that it has a multiplicity that is twice as large as that of *p3*. Incidentally, *p3m1* is the plane group of the LDOS of graphite. One might therefore be inclined to assume that an $F_{16}CoPc$ molecule array is altogether missing in this particular sample area.

STM images of HOPG possess, however, only a few Fourier coefficients so that the LDOS distribution in Fig. 4a is *very* unlikely to be due to a lack of adsorbed $F_{16}CoPc$ molecules. Also the clearly visible vacancies at the periodic motif level in Fig. 4a are straightforwardly explained by the presence of an $F_{16}CoPc$ molecule array.

Table 2: Fourier coefficient residuals for the hexagonal plane groups and the higher symmetric oblique group.

	p6mm	***p6***	***p31m***	***p3m1***	***p3***	***p2***
A_{res} [%]	51.4	50.3	51.4	51.4	50.3	not defined
φ_{res} [°]	30.1	29.9	27.7	16.6	13.6	19.6

Primitive lattice vectors of almost equal length also indicate that a centered unit cell (which would be rectangular and contain two lattice points rather than one) is a reasonable alternative choice. The *a*/*b* ratio of the length of the lattice vectors of such an alternative (centered rectangular) unit cell is approximately 0.622, Fig. 4c, where *a* is in the direction of *x* downwards from the top left corner and *b* is in the direction of *y* to the right as in ref. [6].

Table 3 shows the residuals of the rectangular groups that are type I subgroups of *p3m1* and their respective type I subgroups. The Ao/Ae ratios for the *cm* and *pg* groups are all zero (and this ratio is not defined for the *pm* groups). It is quite remarkable that those rectangular type I subgroups of *p3m1* which have mirror or glide lines oriented nearly perpendicular to the x-axis as shown in the raw STM image (Fig. 4a), i.e. that are labeled as *c1m1*, *p1g1*, and *p1m1*, all possess much smaller phase residuals than their counterparts with orientations nearly parallel to this axis. This suggests that the plane group of the $F_{16}CoPc$ molecule array is indeed *c1m1*. As in Fig. 3d, the site symmetry of the molecules is *m*, i.e. an oblique 2D projection point symmetry of an undisturbed phthalocyanine, see Fig. 4c. Two glide lines are drawn into this figure by their standard symbol as given in ref. [6].

From this and the good match between the hexagonal unit cell and (the two times larger) centered unit cell, we speculate again that the $F_{16}CoPc$ molecules might be inclined with respect to the HOPG surface at cryogenic temperatures. For a comparison of the CIP processed data with classical Fourier filtered data, see Figs. 4d.

Table 3: Fourier coefficient residuals for the rectangular groups that are type I subgroups of *p3m1* and their type I subgroups.

	c11m	*c1m1*	*p11g*	*p1g1*	*p11m*	*p1m1*
A_{res} [%]	13.4	13.4	13.4	13.5	13.3	13.1
φ_{res} [°]	16.6	8.5	17.4	8.7	16.4	8.4

SUMMAY and CONCLUSIONS

CIP was demonstrated on suitable STM images of regular arrays of $F_{16}CoPc$ and CoPc on HOPG as well as CoPc on Ag (111). From these demonstrations, the readers of this paper may infer how large regular 2D arrays of molecules need to be in order to utilize the CIP procedures. The first author of this paper asks the scientific community to send him raw SPM images of 2D periodic arrays so that he can demonstrate the benefits of CIP to a wider audience (in future joint publications). Classical Fourier filtering may be considered as simplest form of CIP, but crystallographic image processing also delivers the symmetry averaged periodic motif and valuable information on the point symmetry of molecules in regular 2D periodic arrays. Also beyond classical Fourier filtering, CIP allows for alternative unit cell choices so that the averaged lattice parameters and the "periodic double motif" contents of centered unit cells can be derived. On the basis of our CIP analyses, we speculate that in the particular cases of the analyzed STM images of our samples (and at cryogenic temperatures), the phthalocyanine molecules might not be lying flat on HOPG.

ACKNOWLEDGMENTS

Portland State University's (PSU's) support in the form of Research Stimulus, Faculty Enhancement, and Internationalization Awards is gratefully acknowledged. PSU's Venture Fund also supported this project. The other members of PSU's Nano-Crystallography Group are thanked for their critical proof-reading of the manuscript. The experimental work at Chemnitz University of Technology was funded by the Deutsche Forschungsgemeinschaft (DFG, International Research Training Group, GRK 1215, "Materials and Concepts for Advanced Interconnects and Nanosystems").

REFERENCES

1. X. Zou and S. Hovmöller, in: *Industrial Applications of Electron Microscopy*, edited by Z.R. Li, Marcel Dekker, New York, 2003, p. 583-614.
2. T. Oku, *Solid State Commun.* **127**, 689 (2003).
3. D. G. Morgan, Q. M. Ramasse, and N. D. Browning, *J. Electron Microsc.* **58**, 223 (2009).
4. P. Moeck, in: *Microscopy: Science Technology, Applications and Education*, Microscopy Book Series, Vol. **4**, edited by A. Méndez-Vilas and J. Diaz, 2010, http://www.formatex.org/microscopy4/.
5. S. Park and C. F. Quate, *J. Appl. Phys.* **62**, 312 (1987).
6. T. Hahn (editor), *International Tables for Crystallography. Brief Teaching Edition of Volume A, Space-group Symmetry*, 5th revised edition, Chester, International Union of Crystallography, 2005.
7. K. Kanatani, *Inter. J. of Computer Vision* **26**, 171 (1998).
8. M. Hietschold, M. Lackinger, S. Griessl, W. M. Heckl, T. G. Gopakumar, and G. W. Flynn, *Microelec. Engin.* **82**, 207 (2005).
9. P. Curie, *J. de Physique* **3**, 393 (1894); see, e.g., the geometric-structural crystallography textbook by S. M. Allen and E. L. Thomas, *The Structure of Materials*, John Wiley & Sons, New York, 1999.
10. T. Kataoka, H. Fukagawa, S. Hosoumi, K. Nebashi. K. Sakamoto, and N. Ueno, *Chem. Phys. Lett.* **451**, 43 (2008).

Mater. Res. Soc. Symp. Proc. Vol. 1318 © 2011 Materials Research Society
DOI: 10.1557/opl.2011.162

Scanning Thermal Lithography as a Tool for Highly Localized Nanoscale Chemical Surface Functionalization.

Joost Duvigneau,[1] Holger Schönherr[2] and G. Julius Vancso[1]
[1] Department of Materials Science and Technology of Polymers, University of Twente, MESA+ Institute for Nanotechnology, Postbus 217, 7500 AE, Enschede, The Netherlands.
[2] Department of Physical Chemistry I, University of Siegen, Adolf-Reichwein-Str. 2, 57076, Siegen, Germany.

ABSTRACT

We report on Scanning Thermal Lithography (SThL), a recently introduced lithographic tool, for local thermochemistry on *tert*-butyl acrylate based polymer films featuring chemical cross links. The tailored polymer films afford platforms for controlled high molecular density coupling and surface immobilization of biologically relevant molecules, such as proteins. The thermally labile *tert*-butyl ester groups in *tert*-butyl acrylate based polymer films can be cleaved in air at temperatures above 150 °C to yield carboxylic acid functional groups for further (bio)-conjugation. The films were optimized to avoid plastic deformation at the elevated temperatures used during SThL. Exploiting these properties patterns with length scales as small as 35 ± 6 nm have been successfully thermally activated with SThL. Hence SThL comprises an attractive approach for the development of *e.g.* (bio)sensors and platforms for cell surface interaction studies with nanoscale patterns.

INTRODUCTION

Recently two exciting, novel approaches, *i.e.* Scanning Thermal Lithography (SThL) and Reactive Imprint Lithography (RIL), have been introduced for the thermochemical surface functionalization of soft condensed matter. RIL has been introduced by us in an effort to exploit heat for combined imprinting (*i.e.* topographical shaping) and chemical functionalization of polystyrene-*block*-poly(*tert*-butyl acrylate) (PS-*b*-P*t*BA) polymer films on microscopic and macroscopic length scales. SThL has been introduced by King and coworkers[1] and by us[2] as an alternative approach for highly localized thermochemical surface functionalization. In SThL heatable cantilevers which comprise resistive heater elements embedded at the cantilever end are utilized for the confined delivery of heat at the probe tip-sample interface (Figure 1). Compared to the well known limitations of conventional lithography or various drawbacks of other scanning probe lithographies (SPLs) heat transport at small length scales is expected to be much less restricted. Since its introduction SThL has proven to be a promising nanotechnology tool for nanoscale chemical surface functionalization.

The reaction mechanism for the thermal deprotection reaction of the *tert*-butyl ester was described elsewhere for PS-*b*-P*t*BA block copolymer film platforms (Figure 1).[2, 3] Upon heating P*t*BA containing films above the thermal deprotection temperature of the *tert*-butyl ester the ester bond is thermally cleaved resulting in the formation of carboxylic acid groups and the liberation of gaseous isobutylene. Upon further heating, anhydride groups are formed through an intramolecular cyclization reaction. Both carboxylic acid and anhydride groups are versatile functionalities for following (bio)conjugation reactions.

Despite this interesting chemistry, the block copolymer films showed some disadvantages. SThL on the PS-*b*-P*t*BA films resulted in the formation of undesired rims of material due to thermomechanical deformations. Additionally (nanoscale) ripple formation was also observed by King and coworker.[1] The design of tailored cross-linked polymer films was hence expected to improve the thermomechanical properties of the films during SThL. Here we introduce a *tert*-butyl acrylate based polymer film platform with improved thermomechanical properties for utilization in SThL and subsequent wet chemical grafting reactions.

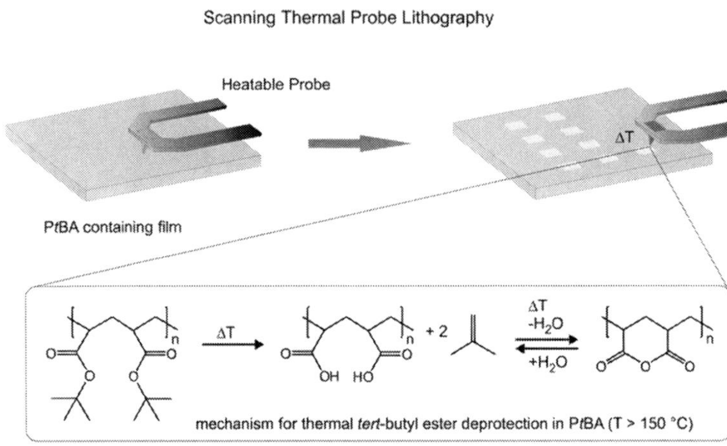

Figure 1. Schematic of SThL on polymer film (top) and the reaction mechanism for thermochemical deprotection of the *tert*-butyl ester protected carboxylic acid groups (bottom). The intramolecular cyclization step is limited in yield to 87 % due to statistical reasons.[4]

EXPERIMENTAL DETAILS

All chemicals were used as received. Monomer mixtures comprising 80 wt% *tert*-butyl acrylate (*t*BA) and 20wt% ethylene glycol dimethacrylate (EGDMA) (which functions as a cross linker) were prepared in a glass vial prior to polymerization. Both monomers were obtained from Sigma Aldrich (St Louis, MO, USA). To these mixtures 3 wt% of the photo initiator Irgacure 184 (a generous gift from Ciba NV, Maastricht, The Netherlands) was added. Subsequently, the monomer mixture was purged with nitrogen. About 0.10 mL of the monomer solution was deposited in the middle of a piranha cleaned 26×76 mm^2 microscopy glass slide (Menzel-Glaser, Braunschweig, Germany). The cleaning comprised the immersion in piranha solution (1:3 (v/v) solution of 30% H_2O_2 and concentrated H_2SO_4) for 10 minutes, rinsing with copious amounts of Milli-Q water and ethanol followed by spin drying, prior to use. ***Caution:*** *Piranha solution should be handled with extreme caution. It has been reported to detonate unexpectedly.*

Subsequently, the solution was covered by placing a microscopy glass slide on top of the first slide. The monomer mixture was allowed to spread between the two slides. Next the samples were irradiated by an array of 6 UV-B lamps (15 W, G15T8E, $\lambda = 306$ nm, Ushio, Tokyo, Japan, sample to lamp distance 5 cm) under nitrogen atmosphere for 30 minutes. Careful separation of the glass slides yielded the approximately 5 µm thick polymer films on one of the microscopy glass slides. Scanning Thermal Lithography: Heatable probes (type AN-2, with 200 µm long cantilevers, Anasys Instruments, Santa Barbara, CA, USA) mounted in a Dimension D3100 atomic force microscope equipped with a hybrid scanner and a NanoScope IVa controller (Bruker/Veeco/Digital Instruments (DI), Santa Barbara, CA, USA) were used for SThL. During the experiments the probe tip temperature (T_i) was controlled with a Nano-TA2 controller (Anasys Instruments). Prior to the experiments, the probe tip temperature was calibrated using the polymer melting point method.[2]

Following SThL of polymer films the in situ formed carboxylic acid groups were activated with 1-ethyl-3-(3-dimethylaminopropyl)carbodiimide hydrochloride (EDC) and N-hydroxy-succinimide (NHS) (obtained from Sigma-Aldrich, St Louis, MO, USA) (200 mM and 250 mM, respectively in PBS). After 1 hour EDC/NHS activation the samples were rinsed 3 times with PBS (B. Braun, Melsungen, Germany) and immersed in a fluoresceinamine (Sigma-Aldrich, St Louis, MO, USA) solution in PBS (concentration ~ 1×10^{-4} M) for 3 hours. Subsequently the samples were rinsed with copious amounts of Milli-Q water and dried under a nitrogen stream. Fluorescence microscopy images of dried fluoresceinamine modified samples were taken immediately after preparation of the samples with an Olypus IX71 fluorescence microscope. For the excitation of fluoresceinamine an U-MWB-2 filter cube was used, typical sample exposure times were 30 ms.

DISCUSSION

Cross linked polymer films exposing *tert*-butyl ester groups were prepared *via* the UV initiated free radical polymerization of a few drops of the corresponding *t*BA and EGDMA monomer mixtures (80 wt% tBA and 20 wt% EGDMA) placed between microscopy cover slides. Monomer to polymer conversion was confirmed with FTIR spectroscopy. The near complete disappearance of the $\nu_{C=C}$ (vinyl) absorbances near 1630 cm^{-1} indicated an almost 100% conversion. The prepared films are from this point on referred to as x-P*t*BA. The conversion of the *tert*-butyl ester to carboxylic acid and anhydride moieties during thermolysis of the x-P*t*BA films at ~ 250 °C for ~ 5 minutes was also confirmed with FTIR spectroscopy.

Scanning thermal lithography on x-P*t*BA films

Following the preparation of x-P*t*BA films, the performance of the x-P*t*BA films during SThL was investigated. Figure 2 shows a contact mode AFM height image with the corresponding cross sectional analysis of an array of 1×1 µm^2 squares written with SThL (T_i ~ 250 °C) as a function of the tip velocity. The reduced apparent height observed for the squares that were written with SThL indicates conversion of the *tert*-butyl ester to carboxylic acid groups (loss of isobutylene). Furthermore upon increasing the writing speed the resulting squares were less deep and less broad. This was also expected for the lower probe tip contact times (*i.e.* reaction time) corresponding to the higher tip velocities. The squares prepared with SThL showed no rim formation at their borders. Hence the cross linked x-P*t*BA films possess improved thermo-mechanical stability compared to the block copolymers studied previously.[2] In addition, SThL at fast scan rates of *e.g.* 40 μm s^{-1} afforded squares that matched the scanned 1 × 1 um^2 area perfectly (*i.e.* no broadening was observed). Hence, it was shown that SThL provides an avenue for the controlled nanometer scale thermochemical patterning of soft condensed matter.

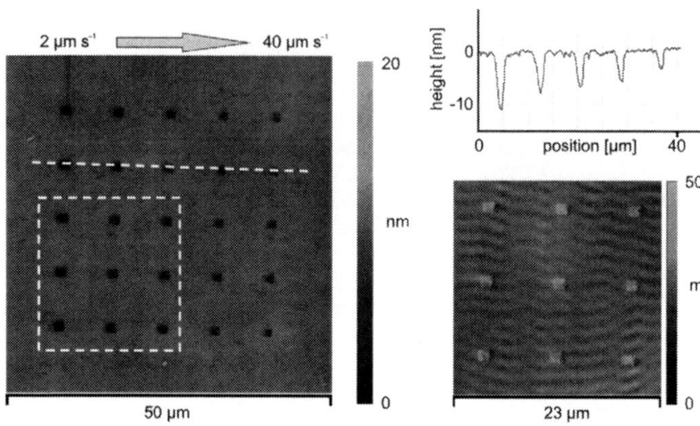

Figure 2. SThL on x-P*t*BA films (scanned area: 1 × 1 μm^2) with increasing tip velocities from left to right. A T_i of ~ 250 °C and a contact load of ~ 20 nN were used. From top to bottom the same SThL parameters were used. The cross section along the white dashed line is shown on the top right, whereas the corresponding friction force image of the squares within the white dashed box is shown at the bottom right.

In addition, Figure 2 shows a friction force microscopy image (zoomed), in which 9 of the 1 × 1 μm^2 squares are visible. Their increased lateral force signal points towards the formation of a more hydrophilic surface area after SThL. This is in agreement with the formation of carboxylic acid groups upon thermolysis.

Chemical derivatization of deprotected areas with fluoresceinamine

The functionality of locally deprotected x-P*t*BA films was confirmed by fluorescence microscopy following wet chemical grafting of fluoresceinamine *via* well-known EDC/NHS chemistry routes. Figure 3 shows a fluorescence microscopy image of a set of 15 × 15 μm^2 squares deprotected by SThL (SThL conditions: T_i ~ 260 °C, velocity 15 μm s^{-1}, load ~ 10 nN). The low back ground emission in the area surrounding the prepared squares points towards the highly localized character of thermochemical surface functionalization by SThL.

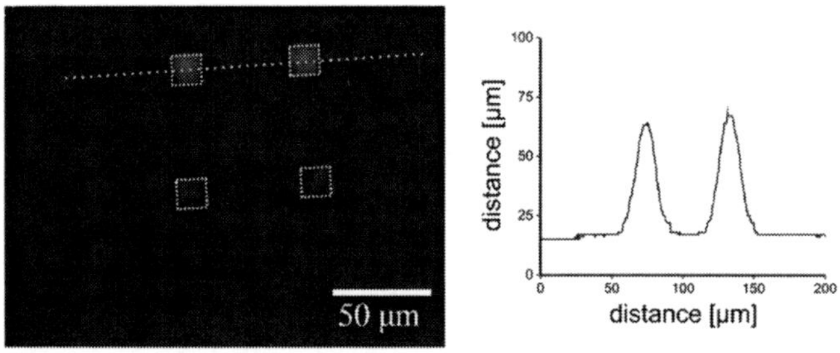

Figure 3. Left: Fluorescence microscopy images of 15 × 15 µm^2 squares (indicated by the white dashed boxes) conjugated with fluoresceinamine following localized thermochemical functionalization with SThL (upper two: T_i ~ 260 °C, load ~10 nN, velocity 15 µm s^{-1}, for the lower two a load of ~50 nN was used). Right: corresponding cross section along the white dashed line in the fluorescence microscopy image is shown.

The lower two squares were written with a higher contact load for which we expected a higher conversion of the *tert*-butyl ester to carboxylic acid groups. However, the observed lower fluorescence emission points towards a less efficient thermochemical conversion to carboxylic acid groups. This is (possibly) explained by *i*) through the increased contact load, the probe tip sample interface increases which results in a decrease in effective T_i, *ii*) depolymerization or *iii*) decarboxylation. However more research is needed to provide conclusive evidence for either one or a combination of the mentioned explanations.

Nanoscale scanning thermal lithography on x-P*t*BA films

Following the introduction of the tailored *tert*-butyl acrylate based polymer films with improved thermomechanical stability, the resolution limit was determined. The smallest groove width obtained so far with SThL on x-P*t*BA films was 35 ± 6 nm as is shown in Figure 4. This number is in good agreement with results reported by King and coworkers for SThL on custom designed polymer films.[5]

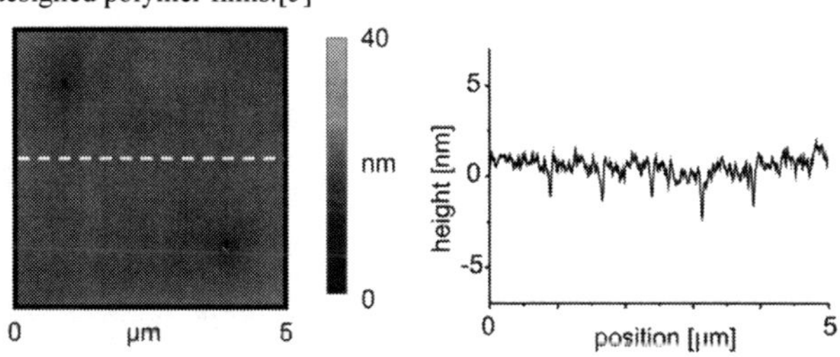

Figure 4. Left: Contact mode AFM height image of a set of grooves written in a x-P*t*BA film by SThL (T_i ~250 °C, load ~ 40 nN, velocity 5 µm s^{-1}). Right: Corresponding cross section along the white dashed line in the AFM height image is shown.

Interestingly the start and end points of the tip trajectory showed deeper depressions with a broadened radial profile, which is consistent with the longer SThL tip residence times at these specific positions (*i.e.* due to the experimental time lack in *i*) controlling heating of the probe and *ii*) moving the probe).These lines were written with a probe with a tip with an estimated radius of curvature of approximately 40 nm. It is expected that future experiments with further optimized settings (*i.e.* sharper tips) will afford patterns with smaller groove widths due to the fact that typically the tip-sample contact radius is smaller compared to the tip radius of curvature. Designing probes with for example multi-walled carbon nanotube functionalized tips (with possible tip radii of ~10 nm) is expected to even further enhance the resolution of SThL.[6] In addition, the use of such probe tips will eliminate tip wear effects (*i.e.* commonly known phenomenon for silicon tips)[7] upon prolonged scanning; hence patterning reproducibility over longer terms will be improved.

CONCLUSIONS

We introduced *tert*-butyl acrylate containing cross linked polymer films as promising versatile platforms for utilization of nanoscale SThL and subsequent wet chemical grafting reactions. Future research is directed towards the systematic investigation of the polymer film composition and to further improving the patterning resolution than the achieved 35 nm line widths. Overall it was demonstrated that SThL is an attractive, promising approach for the controlled nanoscale patterning of polymer surfaces.

ACKNOWLEDGMENTS

This work was supported by NanoImpuls / NanoNed, the nanotechnology program of the Dutch Ministry of Economic Affairs (Grants TPC.6940 and TTF.7217).

REFERENCES

1. R. Szoszkiewicz, T. Okada, S. C. Jones, T. D. Li, W. P. King, S. R. Marder, E. Riedo, *Nano Lett.* **7**, 1064 (2007).
2. J. Duvigneau, H. Schönherr, G. J. Vancso, *Langmuir* **24**, 10825 (2008).
3. J. Duvigneau, S. Cornelissen, N. B. Valls, H. Schönherr, G. J. Vancso, *Adv. Funct. Mater.* **20**, 460 (2010).
4. G. Odian, *Principles of Polymerization*, 3rd ed. (John Wiley & Sons Inc., 1991), p. 692.
5. D. B. Wang, V. K. Kodali, W. D. Underwood, J. E. Jarvholm, T. Okada, S. C. Jones, M. Rumi, Z. T. Dai, W. P. King, S. R. Marder, J. E. Curtis, E. Riedo, *Adv. Funct. Mater.* **19**, 3696 (2009).
6. M. A. Lantz, B. Gotsmann, U. T. Dürig, P. Vettiger, Y. Nakayama, T. Shimizu, H. Tokumoto, *Appl. Phys. Lett.* **83**, 1266 (2003).
7. P. C. Fletcher, J. R. Felts, Z. T. Dai, T. D. Jacobs, H. J. Zeng, W. Lee, P. E. Sheehan, J. A. Carlisle, R. W. Carpick, W. P. *ACS Nano* **4**, 3338 (2010).

Mater. Res. Soc. Symp. Proc. Vol. 1318 © 2011 Materials Research Society
DOI: 10.1557/opl.2011.279

Digital Pulsed Force Mode AFM and Confocal Raman Microscopy in Drug-Eluting Coatings Research

G. Haugstad[1] and K. Wormuth[2]
[1]Characterization Facility, University of Minnesota, Minneapolis, MN, U.S.A.
[2]Surmodics, Eden Prairie, MN, U.S.A.

ABSTRACT

Controlled release of amorphous drug from a polymer matrix depends intimately upon the degree of mixing of drug and polymer, the susceptibility of the drug to crystallization, and the ability of the drug to dissolve and diffuse through water-swollen polymer. Characterization methods ideally would follow these processes on the molecular level in situ and in real time. We move closer to this ideal state of characterization through application of two imaging methods: digital pulsed force mode atomic force microscopy (D-PFM AFM) and confocal Raman microscopy (CRM). We examine model spin-coated films ~1 μm thick containing the drug dexamethasone dispersed in poly(n-alkyl methacrylate) homopolymer and blend coatings. We report aqueous-immersion studies of surface and subsurface structural changes due to drug elution over time frames ranging from very fast (a few minutes) to slow (tens of hours).

INTRODUCTION

One particularly successful application of amorphous drug/polymer mixtures is as coatings on drug eluting stents. Controlled release of an anti-proliferative drug from the coating minimizes restenosis after a stent is placed to prop open a blocked artery. Despite successful application, the nature of amorphous drug elution from stent coatings remains incompletely understood. Open questions revolve around potential crystallization of the drug upon exposure to body fluids, diffusion of water and salts into the coating, and migration of drug out of the coating. Upon exposure of amorphous drug/polymer mixtures to an aqueous medium, water likely plasticizes the drug/polymer mixture and solubilizes the drug. Upon plasticization, drug mobility increases and the probability for drug crystallization rises. However, the presence of polymer may arrest crystallization or kinetically stabilize amorphous drug [1,2]. Here we directly examine drug-polymer coatings immersed in water and watch drug elution and crystallization employing atomic force microscopy for high-resolution (~1-10 nm) surface sensitivity and confocal Raman microscopy for lower resolution (~250-500 nm) but with 3D spatial sensitivity.

EXPERIMENT

Coatings of drug/polymer mixtures containing dexamethasone drug in (a) poly(butyl methacrylate) (PBMA),a 50:50 ratio, or (b) PBMA/poly(lauryl methacrylate) (PLMA) blends, a 43.5:43.5:13 ratio (minority PLMA, of rubbery character), were prepared by spin coating (50 mg solids per ml THF solvent) at 3000 rpm on silicon wafers (native oxide surface) with no further thermal or chemical treatment. Coated wafers were cleaved to provide pieces approximately 2x2-cm in size, large enough to comprise the floor of an open AFM liquid cell.

An Agilent Technologies 5500 scanning probe microscope with an attached digital pulsed force mode controller (D-PFM, WITec, GmbH) allowed the surfaces of the coatings to be probed under water immersion. The integrated environmental sample chamber was maintained at a relative humidity of >80% to prevent evaporation from the liquid cell, which would change the concentration of the eluted drug. (Based on the total drug content in the film, change of film thickness due to elution, the volume of the water medium, and the known solubility of dexamethasone in water, the concentration of released drug was determined to remain well below liquid-cell saturation throughout the experiment.) At each location the D-PFM controller measures the entire force-Z curve to examine tip-sample interaction. This distance cycling was performed sinusoidally in time at a frequency of 2000 cycles per second, using Z amplitudes of a few tens of nanometers, employing silicon tip/cantilevers of ~3 N/m spring constant and ~75 kHz resonant frequency (AppNano). Real-time analysis of the force curves provided images derived from tip-sample adhesion, contact stiffness and surface topography. The latter was obtained via feedback-driven Z scanner displacement to keep the maximum cantilever deflection within each cycle (force signal relative to that out of contact) constant, site to site. Thus the technique is *quasistatic* as in contact mode AFM (force equilibrium) yet involves a brief (~0.1 ms) intermittent contact during which lateral stresses and sliding are minimal.

A WITec Alpha 300 confocal Raman microscope with 532 nm wavelength laser light and a water immersion objective provided Raman images, an entire Raman spectrum collected at each pixel. The images were deconvolved using an augmented classical least squares algorithm yielding images contrasting the spatial distribution of drug polymorphs and polymers.

RESULTS AND DISCUSSION

Raman microscopy

We determined the spatial distribution of PBMA, PLMA, and amorphous and crystalline dexamethasone within coatings by exploiting the significant Raman spectral differences between the polymers and drug polymorphs. In control experiments with confocal Raman imaging, thin coatings of dexamethasone in poly(butyl methacrylate) (PBMA) appear highly resistant to crystallization upon exposure to water. With the presence of a minority fraction of poly(lauryl methacrylate) (PLMA) in the coating, however, one hour of exposure to water produces stark effects. In Figure 1, Raman images show nucleation of dexamethasone crystals primarily in the proximity of PLMA domains, depth resolved to within ~0.5 μm of the surface (roughly half the coating thickness).

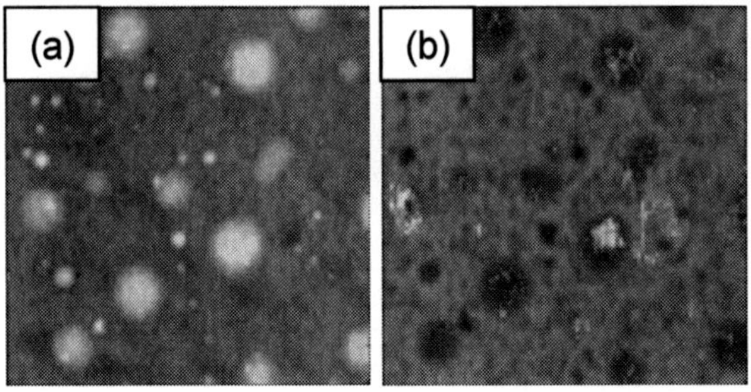

Figure 1. 25x25-μm Raman spectroscopy images of a coating of dexamethasone/PBMA/PLMA under water after exposure to water for 1 hour. (a) Red (dull grey) denotes PBMA, green (bright) denotes PLMA. (b) Red (dull grey) denotes amorphous dexamethasone, green (bright) denotes crystalline dexamethasone; darkest is absence of dexamethasone.

AFM, dexamethasone in PBMA

Raman images of a dexamethasone/PBMA coating (not presented here) show homogeneous mixing (within ~270 nm lateral resolution, i.e., half wavelength of the laser light). In contrast, the representative set of AFM images in Figure 2 reveals a nano-structured surface topography which dramatically evolves upon exposure to water. Prior to exposure to water, the surface contains small, mainly circular bumps ≈200 nm in diameter and ≈20 nm tall (a, bright regions); these also exhibit reduced adhesion with the AFM tip (not shown). Stiffness and adhesion images acquired on dry coatings at elevated temperatures at which PBMA softens (not shown) imply that the bump locations are dexamethasone-rich and the intervening locations polymer-rich. After approximately 10 minutes of exposure to water, a significant number of circular pits or pores appear in place of bumps (b, dark regions), which suggests that a first subpopulation of bumps rapidly releases drug. A second, lesser subpopulation of bumps bulge out vertically by as much as 50 nm (and are apparently "sharp" enough to partially reverse-image the pyramidal AFM tip shape). After 150 minutes exposure to water (c), the initial holes further deepen while remaining bumps decrease in height. After 23 hours in water (d), holes have further expanded while bumps have receded to near the "baseline" of the coating between the holes.

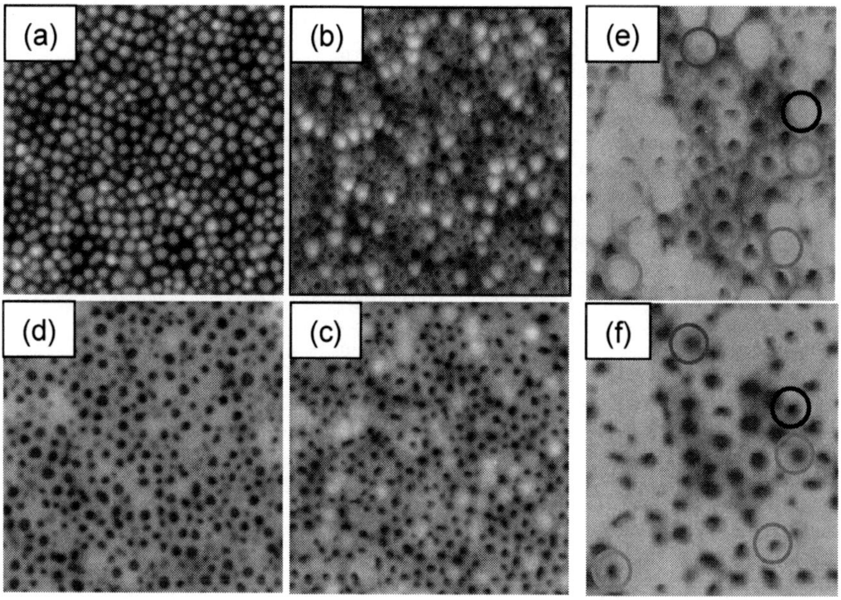

Figure 2. (a-d) Representative 5x5-μm height images of dexamethasone/PBMA coating in initial dry state (a) and at three time stages under water: (b) ~10 minutes, (c) 150 minutes, (d) 23 hours. (e-f) 2.2x2.4-μm images, same time stages as (b-c). Circles locate protrusions that become holes.

Closer inspection of images of dexamethasone/PBMA reveals that a third, even lesser subpopulation of bumps observed during the first few minutes of water exposure that convert to holes after a few tens of minutes. Five examples are denoted with circles in Figure 2(e-f). This suggests the presence of a very thin skin of PBMA over some ~100-nm drug domains; the skin may later rupture to enable drug elution, whereas the first subpopulation of bumps seem to suddenly release drug upon water exposure, as if there is either no PBMA skin or a PBMA skin that is immediately perforated. Finally, the vertical growth then shrinkage of the second subpopulation of protrusions suggests PBMA-encapsulated drug domains that are penetrated and expanded by water faster than drug can diffuse out in three dimensions; but ultimately the latter process causes the protrusions to collapse. Characterized over a full day, the vast majority of circular, drug-rich domains either converted to holes immediately or expanded then contracted to baseline. Further studies could further elucidate these behaviors. We anticipate that there is a continuum of behavior, with a range of drug elution times best represented on a logarithmic scale, rather than three distinct rates as suggested above for expository purposes.

AFM, dexamethasone in PBMA/PLMA blend

AFM images of coatings containing the two-polymer blend exhibit rich morphologies spanning nano- to micro-scales. Figure 3 contains images of height and tip-sample adhesion (pull-off force) on a dexamethasone/PBMA/PLMA coating. Similar to the dexamethasone-PBMA case there are small, nearly circular bumps ≈200 nm in diameter, which exhibit relatively low adhesion with the AFM tip. But unlike the dexamethasone/PBMA case there are intermediate-size domains (≈500-nm diameter) that exhibit high adhesion, plus large domains (diameters ranging from 1-4 μm as compiled in 50-μm survey images) that are either similar in surface height to the intermediate domains or bowl-shaped depressions, tens of nanometers and in some cases more than 100-nm deep. The large domains exhibit adhesion that is high compared to the small bumps but not quite as high as the intermediate-sized domains. Both the intermediate and large domains exhibit equally low contact stiffness (not shown), indicating primarily PLMA content, as this polymer is rubbery at room temperature, whereas PBMA is glassy. Comparison with Raman micrographs as in Figure 1 confirms this assignment for the large domains. AFM images on drugless (control) PBMA-PLMA coatings (not shown) also contain both the intermediate and large-sized PLMA domains, and with similar slight differences in adhesion, but in both cases lower than the adhesion measured on PBMA for similar maximum loading forces (~10 nN; whereas PLMA exhibits greater adhesion for substantially larger pushing forces at turnaround, i.e., greater viscoelastic adhesion memory).

Upon immersion in water, dramatic changes in both morphology and adhesion are immediately apparent. As with the simpler dexamethasone/PBMA coatings, many of the small bumps are replaced by small holes on the ~10-minute time frame, then grow in depth over times spanning hours. Another similarity is the outward bulging of a small subpopulation of the original small bumps. The intermediate-sized, PLMA-rich domains more significantly bulge outwards, by ≈200 nm, suggesting local penetration of water and/or subsurface aggregation of drug; whereas the large PLMA domains seemingly remain unchanged in relative surface height. With passing time the adhesion signatures also evolve, generally towards lower adhesion, except for the large PLMA domains which remain remarkably constant in adhesion.

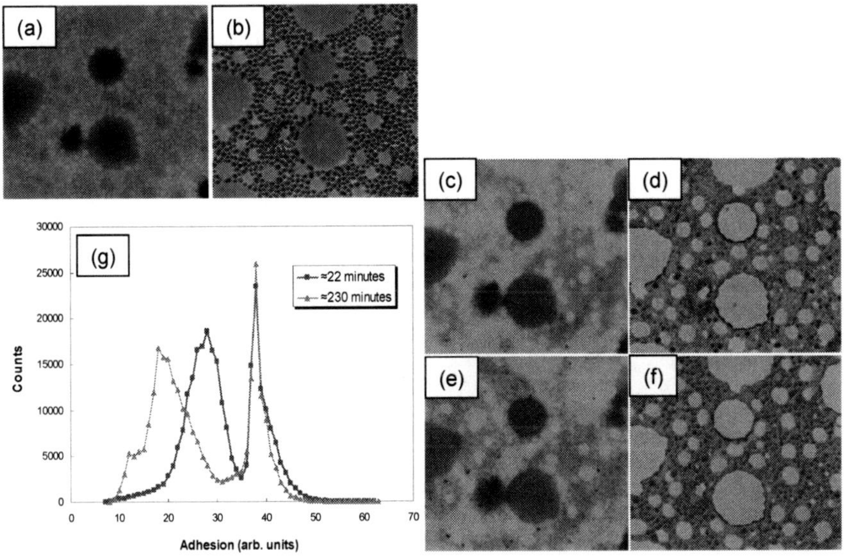

Figure 3. (a-b) Representative 10x9-µm AFM height (a) and tip-sample adhesion (b) images of dexamethasone-in-PBMA/PLMA coating in its initial dry state. (c-f) Representative 10x10-µm AFM height (c,e) and tip-sample adhesion (d,f) images of the same upon immersion in water for 22 (c-d) and 230 (e-f) minutes. (g) Histograms of the adhesion images (d,f).

These latter observations are quantified as histograms of adhesion images, presented in Figure 3g. The tall, sharp peak is the signal derived from the large PLMA domains. Its constancy indicates a lack of surface changes within these domains, which at first seems inconsistent with the appearance of a strong crystallized drug signal observed in the immersion Raman images (Figure 1). Thus we interpret this crystallization Raman signal as originating at depth, whether buried inside the PLMA domains or below them; but most likely at their peripheries given that Raman experiments show strong preferential partitioning of dexamethasone into PBMA, with little dexamethasone signal found in PLMA domains. We suggest that it is the presence of interfaces with PLMA that triggers the crystallization of dexamethasone.

The reduction of adhesion with passing time seen in Figure 3g for all other surface constituents is intriguing. The adhesion measurement is fundamentally a probe of contact hysteresis, in turn a gauge of molecular mobility during the formation and break of tip-sample contact. Molecular rearrangements during contact allow the interfacial system to achieve a lower-energy state that requires a pulling force to disrupt [3]. Our observations suggest a decrease in surface molecular mobility as drug is eluting from the coating. Certainly the dexamethasone molecules are mobile, and as they evacuate from the near-surface region one might expect a decrease in overall molecular mobility of the matrix; this is especially likely because the presence of dexamethasone in PBMA should plasticize the polymer. The shifting of what appear to be at least three distinct surface components – both the low adhesion, broad peak and shoulder, and a high-adhesion shoulder – suggest that the removal of dexamethasone reduces

PBMA conformational mobility, and perhaps PLMA mobility as well (in the ~500-nm domains, perhaps near the PLMA-PBMA interface). Control experiments on drugless coatings show no changes in PBMA or PLMA during immersion.

CONCLUSIONS

Digital pulsed force mode AFM and confocal Raman microscopy were employed, together with water immersion, to elucidate the structure and behavior of ~1-μm thick films of dexamethasone/PBMA (50:50) and dexamethasone/PBMA/PLMA (43.5:43.5:13). AFM findings include (a) the rapid formation (burst) of ~200-nm diameter nanoscale surface holes in place of a majority subpopulation of ~200-nm diameter bumps present in dry, as-cast films, due to rapid elution of dexamethasone; (b) the slower appearance of a small number of additional holes in place of the original bumps, and the growth then collapse of the larger number of remaining original bumps over a period of many hours, due to much slower elution of dexamethasone. The presence of an unbreached PBMA surface barrier is proposed to result in a slower, diffusional release of drug (process b), whereas molecular-scale breaks in this barrier (likely below AFM resolution) may allow for rapid evacuation of dexamethasone from pockets (process a). Raman microscopy findings include the immersion-induced formation of dexamethasone microcrystallites in the vicinity of PLMA. The lack of surface crystallites in AFM images indicates that Raman is sensing objects below the surface, likely interfacing with PLMA domains. We conclude that complementary D-PFM AFM and confocal Raman microscopy provide insights into the mechanisms of drug elution from, and crystallization within, biomedical coatings. Revealed relationships between coating structure and release kinetics (not presented here) are useful to the development of drug-eluting coatings formulations.

ACKNOWLEDGMENTS

Support was provided by Surmodics, Inc., via its membership in the Industrial Partnership for Research in Interfacial and Materials Engineering at the University of Minnesota. Parts of this work were carried out in the University's Characterization Facility (a Center within the College of Science and Engineering), which received partial support from the NSF through the NNIN and MRSEC programs.

REFERENCES

1. H. Konno and L. Taylor, *J. Pharm. Sci.* **95**, 2692, 2006.
2. P. Marsac, S. Shamblin, L. Taylor, *Pharm. Res.* **23**, 2417, 2006.
3. J. N. Israelachvili, *Intermolecular and Surface Forces*, Academic Press, 1991.

Mater. Res. Soc. Symp. Proc. Vol. 1318 © 2011 Materials Research Society
DOI: 10.1557/opl.2011.280

Measurement of Piezoelectric Transverse and Longitudinal Displacement with Atomic Force Microscopy for PZT Thick Films.

Yuta Kashiwagi[1, 2], Takashi Iijima[2], Toru Aiso[3], Takashi Yamamoto[4], Ken Nishida[4], Hiroshi Funakubo[5], Takashi Nakajima[1] and Soichiro Okamura[1]

[1]Tokyo University of Science, 1-3 Kagurazaka, Shinjuku-ku, Tokyo 162-8601, Japan
[2]National Institute of Advanced Industrial Science and Technology, AIST Tsukuba West, 16-1 Onogawa, Tsukuba, Ibaraki 305-8569, Japan
[3]Toyo Corporation, 1-1-6 Yaesu, Chuo-ku, Tokyo 103-8284, Japan
[4]National Defense Academy, 1-10-20 Hashirimizu, Yokosuka, Kanagawa 239-8686 Japan
[5]Tokyo Institute of Technology, 4259 Nagatsuda, Midori-ku, Yokohama 226-8502, Japan

ABSTRACT

The actual transverse and longitudinal displacement of PZT thick film was measured using a newly developed atomic force microscopy (AFM). The AFM is attached a feedback circuit named "torsion feedback". The torsion and Z-height feedback circuits control an AFM cantilever to follow piezoelectric deformation of the sample. To measure transverse displacement, the cantilever contacts the edge of sample. The transverse displacement is determined from the torsion feedback signal absolutely. To measure longitudinal displacement, the cantilever contacts the center of sample. The longitudinal displacement is determined from Z-height feedback signal absolutely. A 5-μm-thick PZT film was prepared on $Pt/Ti/SiO_2/Si$ substrates. The film sample was shaped square pillar. The side electrode length (L) of square pillar shaped sample was ranged from 1000 μm to 10 μm. The relation between side electrode length and the transverse or the longitudinal displacements were investigated. With decreasing L, the transverse displacement decreased nonlinearly, and the longitudinal displacement increased nonlinearly. The finite element method (FEM) simulation suggests that the substrate clamped PZT film behaved nonlinearly. The effective $-d_{31}$ and d_{33} were calculated from the measured displacement, and these values increase with decreasing L. The effective d_{33} and $-d_{31}$ showed correlation, and the ratio was $d_{33} : -d_{31} = 5.3 : 1$, whereas the bulk ratio is $d_{33} : -d_{31} = 2.4 : 1$. This result suggests that the substrate clamping effect of the transverse displacement was larger than that of the longitudinal displacement.

INTRODUCTION

Piezoelectric materials show the generation of the electric polarization when they are strained and this effect is called direct effect. On the other hand, the piezoelectric materials also show the generation of the strain directly proportional to an applied field and this effect is called converse effect. For poled piezoelectric ceramics, the converse piezoelectric constants, $[d_{ij}]$, is usually treated as 2nd rank matrix because of symmetry of crystal. The d_{31} is the coefficient relating the electric filed along the polar axis to the strain perpendicular to the field, while the d_{33} is corresponding coefficient for both strain and the field along the polar axis. [1][2] For piezoelectric films, the electric field direction (x_3 axis) is defined as perpendicular to the film surface. [3]

In recent years, piezoelectric films have been well investigated to develop micro-actuators for microelectromechanichal systems (MEMS). [4] [5] To design the MEMS micro-actuator devices, the evaluation of the longitudinal and transverse piezoelectric properties of the films is required. Especially, d_{31} is important, because the qualification of the micro-actuators are mainly calculate form the d_{31}. For piezoelectric bulk ceramics, d_{33}, and the d_{31} are determined with various standard such as the resonance method and dynamic load method. On the other hand, standard evaluation methods of d_{33} and d_{31} for the piezoelectric films are not developed, because substrates clamp the piezoelectric films and influence the piezoelectric film deformation. [3] [6] [7] In the case of the longitudinal displacement measurement of the piezoelectric films, some evaluation methods were reported such as a double beam interferometer (DBI) [8] [9] and an atomic force microscopy (AFM) [10] [11]. However, direct evaluation method of transverse displacement for the d_{31} is not so many. Up to now, transverse properties of piezoelectric films were indirectly calculated from a deformation of a cantilever shaped film structure sample. [12] In this study, the actual transverse and longitudinal displacements of PZT film samples were evaluated using feedback AFM system.

EXPERIMENT

A $Pb(Zr,Ti)O_3$ (Zr:Ti=53:47) thick film was deposited on $Pt/Ti/SiO_2/Si$ substrate using a chemical solution deposition (CSD) process. The film thickness was 5 μm. After the film deposition, Pt top electrode was deposited by sputtering. Then square pillar shaped PZT structures were fabricated using photolithography and reactive ion etching process. Details of the fabrication process were described elsewhere. [13] The side length of square pillar shaped sample was defined as L. Prepared L was ranged from 1000 μm to 10 μm. To measure the transverse displacement, a newly developed feedback circuit named "torsion feedback" was attached to the AFM (Nano-R; Toyo Corp). For the transverse displacement measurement, a conductive AFM cantilever contacts the edge of the top electrode and applies electric field to the film. When the piezoelectric film shows the transverse deformation, the cantilever is twisted. Photo-detector detects the torsion of cantilever by reflected laser from the cantilever. The torsion feedback circuit controls the cantilever using X-axis piezo-actuator to follow the transverse displacement. The transverse displacement is determined from the feedback signal absolutely. On the other hand, the AFM cantilever contacts the center of the top electrode to measure the longitudinal displacement. When the cantilever applies electric field and film shows the longitudinal deformation, a usual Z-height feedback circuit controls the cantilever using Z-axis piezo-actuator to follow the longitudinal displacement. The longitudinal displacement was estimated with the Z-height feedback signal absolutely. The AFM system was connected to ferroelectric test system (FCE-1; Toyo Corp.), and the system can measure the polarization and piezoelectric displacement simultaneously. Detail of measurement configuration is exhibited in Fig. 1. The measurement frequency of the applied electric field is fixed at 5 Hz, because limitation of the X- and Z- axis feedback circuit. To align the polar axis and to increase the piezoelectric displacement, the poling process was carried out. A DC field which amplitude was 100 kV/cm was applied to the sample for 6 min. The shape of measuring applied electric field was unipolar triangle shape, and the amplitude of applied electric field was 50 kV/cm.

RESULTS AND DISCUSSION

Figure 2 shows relation between the transverse displacement and electrode length L. The straight line represents an ideal relation between the transverse displacement and L for free-standing (without substrate) PZT films described as follows.

$$\Delta S_{31} = -d_{31}\frac{V}{t_f}L \tag{1}$$

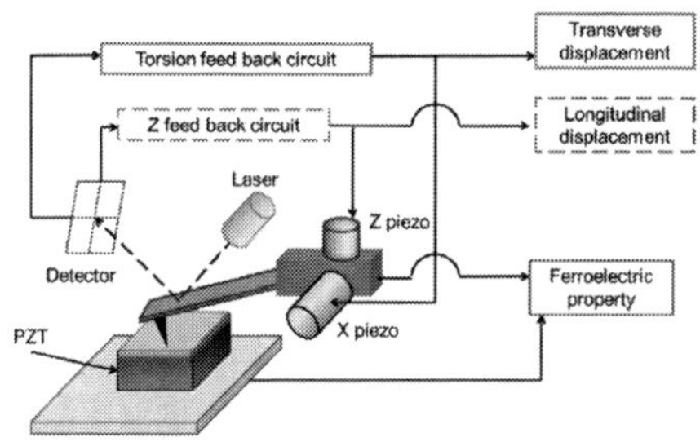

Figure 1. Schematic illustration of the measurement configuration using the feedback AFM system

Where ΔS_{31} is transverse displacement, d_{31} is the transverse piezoelectric constant (d_{31} of bulk PZT is -93.5 pm/V [14]), V is the applied voltage, and t_f is the film thickness. Solid lines of 0.5, 2, and 10μm indicate the FEM simulated relation between the transverse displacement and the L for 0.5-, 2- and 10- μm-thick films, respectively. [15] As shown in Fig. 2, the measured transverse displacements and L show nonlinear relationship, while the ideal transverse displacement show linear relationship. The measured transverse displacement showed the same tendency with FEM simulation. This means that the measured transverse displacements include substrate deformation. From these results, it is concluded that the substrate clamping affected this nonlinear behavior.

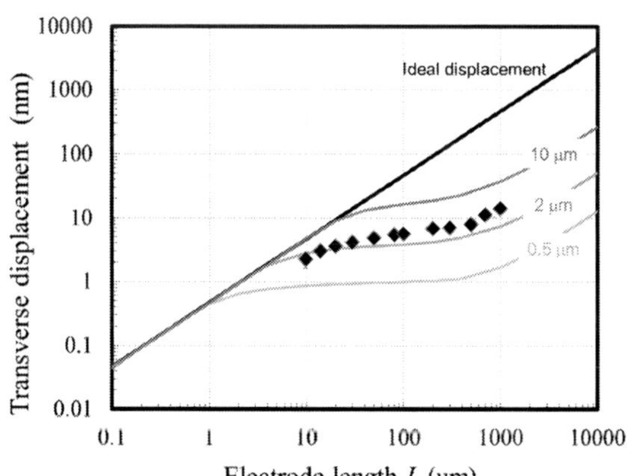

Figure 2. Relation between the transverse displacement and electrode length.

Figure 3 shows relation between the longitudinal displacement and L. The longitudinal displacement increased nonlinearly with decreasing L. When L was more than 100 μm, the longitudinal displacement showed negative amount. This means that the PZT film bended downward with the substrate. [16][17]The straight line indicates ideal longitudinal displacement, which means that of the free-standing PZT films as describes follows.

$$\Delta S_{33} = d_{33} \times V \tag{2}$$

Where ΔS_{33} was longitudinal displacement, d_{33} is the longitudinal piezoelectric constant (d_{33} of bulk PZT is 223 pm /V [14]). Fig. 3 suggests that measured longitudinal displacement approach ideal amount with decreasing electrode length, L. This result indicates that the substrate bending effect decrease with decreasing L. Finally, the substrate bending is considered to vanish, and the measured longitudinal displacement amount becomes equal to the ideal amount when L is less than the film thickness. Figure 4 shows electrode length dependence on the effective d_{33} and the effective $-d_{31}$ calculated from the equation (1) and (2). The straight lines represent the bulk d_{33} and $-d_{31}$ value. Figure shows that the effective d_{33} and the effective $-d_{31}$ increased with decreasing electrode length. This result suggests that the substrate clamping effects decreased with decreasing electrode length. When the electrode length was 10 μm, the measured effective d_{33} value was close to the bulk d_{33} value. However, the measured effective $-d_{31}$ value was smaller than the bulk $-d_{31}$ value. Figure 5 shows relation between the effective d_{33} and the effective $-d_{31}$. Square shaped point represents bulk piezoelectric constants. As shown in Fig. 5, the effective $-d_{31}$ and the effective d_{33} show linear relationship. Therefore, it seems that there is correlation between the effective d_{33} and the effective $-d_{31}$. The calculated correlation coefficient was 0.99. From the slope of effective piezoelectric constants, ratio of effective d_{33} to effective $-d_{31}$ was 5.3 : 1, whereas the bulk amount is $d_{33} : -d_{31} = 2.4 : 1$. This result suggests that the substrate clamping effect of the transverse displacement is larger than that of the longitudinal displacement.

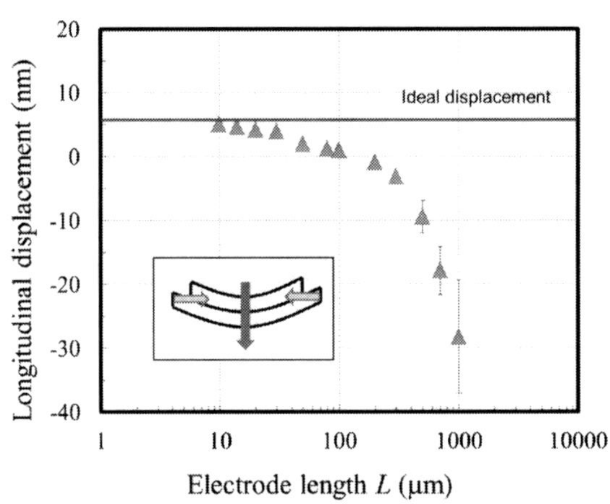

Figure 3. Relation between the longitudinal displacement and electrode length.

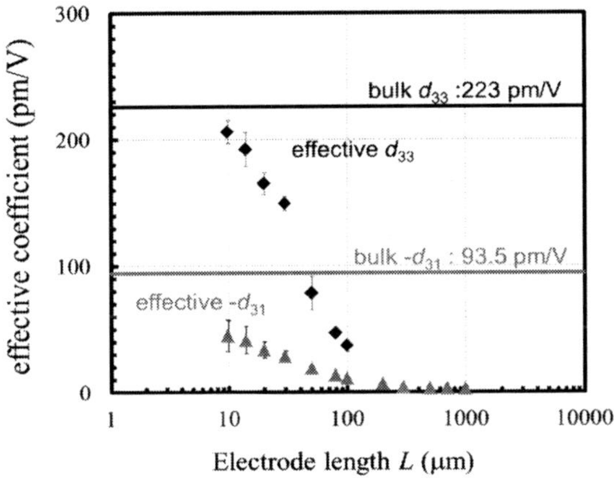

Figure 4. Electrode length dependence for the measured effective d_{33} and effective $-d_{31}$.

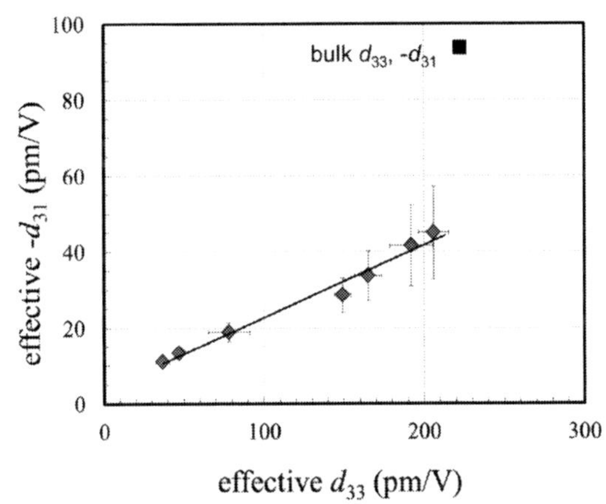

Figure 5. Relation between the measured effective d_{33} and the effective $-d_{31}$.

CONCLUSIONS

The actual transverse and longitudinal displacement of PZT thick film was measured using a newly developed AFM. The torsion and Z-height feedback circuits control an AFM cantilever to follow piezoelectric deformation of the sample. The relation between side electrode length,*L* , and the transverse or the longitudinal displacements were investigated. With decreasing *L*, the transverse displacement decreased and the longitudinal displacement increased nonlinearly. The effective $-d_{31}$ and d_{33} were calculated from the measured displacement, and these values increase with decreasing *L*. This result suggests that the substrate clamping effects decreased with decreasing electrode length because the PZT film deformed with the substrate. The effective d_{33} and $-d_{31}$ showed correlation, and the ratio was $d_{33} : -d_{31} = 5.3 : 1$, whereas the bulk ratio is $d_{33} : -d_{31} = 2.4 : 1$. Therefore, the substrate clamping effect of the transverse displacement was larger than that of the longitudinal displacement.

REFERENCES

1. R. E. Newnham, *Properties of Materials, Anisotropy, Symmetry, Structure*.(Oxford University Press, New York, 2005) p. 87.
2. A. J. Moulson, and J. M. Herbert, *Electroceramics, Materials, Properties, Applications*. 2nd ed. (Wiely, Chichester, 2003) p.344.
3. K. Lefki, and G. J. M. Dormans, *J. Appl. Phys.*, **76**, 1764 (1994).
4. S. T. Mckinstry, and P.Muralt, *J. Electroceram.*, **12**, 7 (2004).
5. N. J. Conway, Z. J. Traina, and S-G. Kim, *J. Micromech. Microeng.*, **17**, 781 (2007).
6. L. Chen, J-H. Li, J. Slutsker, J. Ouyang, and A. L. Roytbourd, *J. Mater. Res.*, **19**, 2853 (2004).
7. A. Barzegar, D. Damjanovic, N. Ledermann, and P. Muralt, *J. Appl. Phys.*, **93**, 4756 (2003).
8. A. L. Kholkin, E. L. Colla, A. K. Tagantsev, D. V. Taylor, and N. Setter, *Appl. Phys. Lett.*, **68**, 2577 (1996).
9. Y. Kashiwagi, T. Iijima, T. Nakajima, and S. Okamura, *J. Ceram. Soc. Jpn.*, **118**, 640 (2010).
10. J. A. Christman, R. R. Woolcott, Jr., A. I. Kingon, and R. J. Nemanich, *Appl. Phys. Lett.*, **73**, 3851 (1998).
11. T. Iijima, S. Ito, H. Matsuda, *Jpn. J. Appl. Phys.*, **41**. 6735 (2002).
12. I. Kanno, H. Kotera, K. Wasa, *Sens. Actuators A*, **107**, 68 (2003).
13. T. Iijima, S. Osone, Y. Shimojo, H.Nagai, *Int. J. Appl. Ceram. Technol.*, **3**, 442 (2006).
14. D. A. Berlincourt, C. Cmolik, and H. Jaffe, *Proc. IRE*, **48**, 220 (1960).
15. T. Yamamoto, M. Yamamoto, K. Nishida, H. Funakubo, T. Iijima, T. Aiso, and Y. Ichikawa, *Jpn, J. Appl. Phys.*, **48**, 09KA04 (2009).
16. T. Iijima, K. Kunii, *Jpn. J. Appl. Phys.*, **40**, 5740 (2001).
17. H.Okino, M. Hayashi, T. Iijima, S. Yokoyama, H. Funakubo, N. Setter, and T.Yamamoto, *Mater. Res. Soc. Symp. Proc.*, **902E**, 49 (2006).

Mater. Res. Soc. Symp. Proc. Vol. 1318 © 2011 Materials Research Society
DOI: 10.1557/opl.2011.281

Circular AFM Mode: A New AFM Mode for Investigating Surface Properties

Olivier Noel[1], Pierre-Emmanuel Mazeran[2], Hussein Nasrallah[1]
[1] Molecular Landscapes & Biophotonics Group, LPEC UMR CNRS 6087, Université du Maine, Le Mans, France
[2] Surface Mechanics Group, Laboratoire Roberval, UMR CNRS 6253, Université de Technologie de Compiègne, France

ABSTRACT

For the first time, a new AFM mode is presented that simultaneously allows the measuring of adhesion and friction forces at different constant and continuous sliding velocities. Our methodology consists of implementing a circular relative displacement of the contact to reach a constant sliding velocity, with no stop periods. Some of the main advantages of performing a circular displacement is that continuous and high sliding velocities (more than 1 mm/s) can be reached compared to the low sliding velocities (up to 10 μm/s) available when using commercial AFM. Also, a stationary state is reached when doing measurements. Moreover, the circular mode can be coupled with the classical operating mode, for instance, force spectrum. Main applications of this circular mode are related to metrological measurements in physics that require high speed displacements. As an example, we report the evolution of friction and adhesive forces measured in air at different high sliding velocities.

INTRODUCTION

The development of new techniques such as the Atomic Force Microscopy (AFM) offers new opportunities to understanding, at a nanometric scale, phenomenological mechanisms encountered in many fields such as material science, mechanics, biology or microelectronics. Indeed, this widely used technique has become a powerful analytical tool for measuring intermolecular interaction forces with high sensitivity. In particular, it is possible to generate a relative back and forth displacement of a nanometer probe to a sample surface. The amplitude of this relative displacement ranges from 0 to 100 μm, with typical frequencies ranging between 0.1 and 10 Hz, while measuring torsion forces (for a displacement of the probe in the plane of the sample) or adhesive forces (for a displacement of the probe perpendicular to the sample) [2]. It also offers the capability of imaging surface properties and topography with high resolution [3].

Although the design of high-speed AFM has improved to get fast surface properties imaging [4-7], it is still challenging to get experimental accurate results for displacement velocities higher than 100 μm/s. In addition, the inversion of the movement direction when performing image scanning or spectra force acquisition generates stop periods during which the nature of the tip/sample contact may evolve significantly. In all cases, no stationary state is obtained during measurements or scanning.

Here, we present for the first time a new AFM mode called "the circular AFM mode"[1] that allows collecting data at different constant and continuous high sliding velocities (higher than 1 mm/s) without any stop periods. Our methodology consists of modifying an AFM to achieve a relative displacement of the probe/sample contact in a circular motion. The sliding velocity is governed by both the radius of the circular motion and its frequency (see the experimental section). Thus, the circular imprints resulting from wear or plastic deformation obtained with the circular AFM mode (Figure 1) clearly show the circular motion of the relative displacement of the probe/sample contact for different sliding velocities.

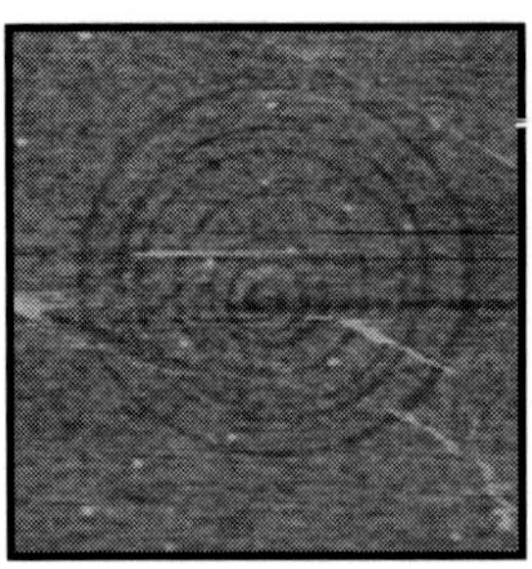

Figure 1: AFM topographic image (7μm × 7μm) of an AsGa thin film surface. Imprints illustrate the wear created by the circular displacement of the probe relative to the sample in using the AFM Circular mode at different sliding velocities.

EXPERIMENTAL

The AFM can be divided in three parts: The probe that senses the interactions with the sample, the 3D scanner that allows the relative displacement of the probe and the optical detector system that allows the measuring of interactions. Usually, the displacement modulus is made of piezoelectric ceramics actuators connected to the probe or the stage. When a voltage is applied to such actuators, they deform themselves in the direction of the applied voltage. For example, voltages applied in both the [-X, X] and [-Y,Y] directions to the ceramics (where X and Y axes refers to the axes defining the plane of the sample) generate a relative displacement of the probe/sample contact in the plane of the sample.

The relative circular displacement of the probe/sample contact requires a specific control of the piezoelectric ceramics where a circular movement is generated by the combination in X and Y of two sinusoidal phase shifted by $\frac{\pi}{2}$ voltages applied to the ceramic actuators. Displacements are generated and controlled by a Digital Analogic Converter (DAC) device, a lock-in amplifier or a sinusoïdal tension generator coupled with a dephasor. If the sample is flat and horizontal, no servo loop is required to maintain a quasi-constant load on the sample.

In such a mode, two parameters of control are avalaible to adjust the velocity of the circular movement. The first one is the voltage amplitude that sets the radius of the circular movement, and the second is the voltage frequency that defines the time needed to make a complete circle. For example, for a full range actuator amplitude of 100 μm corresponding to an applied voltage of 440 V (usual scanner range voltage for commercial AFM), and a frequency of the circular movement of 300 Hz (this frequency is below the resonance frequency of usual commercial actuators), the sliding velocity of the probe is 10^5 μm/s. Thus, sliding velocities ranging from 0 to 10^5 μm/s are easily attainable with usual commercial actuators when using the circular AFM mode.

ADVANTAGES AND APPLICATIONS OF THE CIRCULAR MODE

The relative velocity of the probe/sample contact in the [X, Y] plane is at least three decades higher than conventional velocities attainable with commercial AFM as no servo loop is necessary to control the motion. These velocities are comparable to realistic velocities of displacement of the contacts in nano or microdevices (MEMs, hard disks…). Thus, it is possible to realise local force measurements (at the nanometric scale) with a high resolution and realistic conditions.

In addition, in the classical imaging mode, data are acquired using back and forth displacements. This kind of displacement leads to stoppage of the tip/sample contact each time the sliding direction is inversed. During these stops, the contact evolves from a dynamic contact to a static contact, which leads to a change of the tip/sample contact characteristics during the collecting of the data, like for example, shear stress, adhesion force, and then total applied load. Hence, every experiment conducted by classical AFM modes could be affected by the duration of these stops that depend on the experimental conditions (frequency, scan size, friction force, contact stiffness...). The inconvenience due to stop periods generated by the inversion of the direction during the back and forth movement are eliminated with the AFM circular mode. Thus, a metrological aspect in nanotribology for example or in probing properties at the nanoscale can be considered.

Combined to force spectra mode, it is possible to measure simultaneously adhesion forces and friction forces or friction coefficient while the relative probe/sample displacement is circular (Figure 2). We checked that the probe is not submitted to artefacts like jumping onto the surface during sliding, in doing measurements with probes of different stiffnesses. Moreover, the friction force values on the friction spectra are the amplitude of the torsional displacement of the probe corresponding to the frequency of the circular motion. These values can be easily converted into force units after calibration.

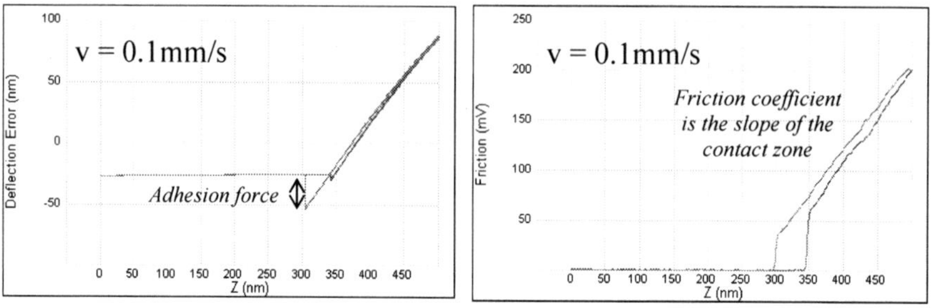

Figure 2: Simultaneous acquisition of adhesion force (left curve) and lateral force (right curve) spectra in using the AFM circular mode.

CONCLUSIONS

The AFM circular mode combines modern techniques that enable the monitoring of force spectra and imaging with a high resolution at the nanoscale while the probe/surface contact is sliding in a circular motion. The major breakthroughs are that the relative probe/surface contact circular velocity is constant and three decades higher than commercial available velocities and, that measurements are done in stationary states, which mean that no stop periods can affect significantly the contact between the probe and the surface. Moreover, the velocity values with the AFM circular mode are much closer to realistic sliding velocities in nano or micro systems than the ones obtained with commercial AFM modes. Applications for the characterisation of surface properties in metrology, nanotribology or adhesion sciences are obvious. This new AFM mode should also be powerful to get new insights in physics, biological or chemical mechanisms.

ACKNOWLEDGMENTS

This work was entirely supported by ANR project number ANR-08-JCJC-0051-01

REFERENCES

[1] O. Noel, P-E. Mazeran, H. Nasrallah, AFM Circular Mode, Patent in 2010.

[2] H.J. Butt, B. Cappella, M. Kappl, Force measurements with the atomic force microscope: Technique, interpretation and applications, Surface Science Reports, 59 (2005) 1-152.

[3] S. Scheuring, Y.F. Dufrene, Atomic force microscopy: probing the spatial organization, interactions and elasticity of microbial cell envelopes at molecular resolution, Molecular Microbiology, 75 (2009) 1327-1336.

[4] S.R. Manalis, S.C. Minne, C.F. Quate, Atomic force microscopy for high speed imaging using cantilevers with an integrated actuator and sensor, Applied Physics Letters, 68 (1996) 871-873.

[5] T.J. McMaster, D. Brayshaw, M.J. Miles, A.E. Walsby, P. Dunton, A new ultra high speed AFM technique for biophysics: 3-diemensional imaging of surfaces, molecules and processes with true millisecond resolution, Biophysical Journal, 88 (2005) 541A-541A.

[6] P. Parot, Y.F. Dufrene, P. Hinterdorfer, C. Le Grimellee, D. Navajas, J.L. Pellequer, S. Scheuring, Past, present and future of atomic force microscopy in life sciences and medicine, Journal of Molecular Recognition, 20 (2007) 418-431.

[7] I. Casuso, N. Kodera, C. Le Grimellec, T. Ando, S. Scheuring, Contact-Mode High-Resolution High-Speed Atomic Force Microscopy Movies of the Purple Membrane, Biophysical Journal, 97 (2009) 1354-1361.

Mater. Res. Soc. Symp. Proc. Vol. 1318 © 2011 Materials Research Society
DOI: 10.1557/opl.2011.282

Scanning Probe Microscopy with Diamond Tip in Tribo-nanolithography

Oleg Lysenko[1], Vladimir Grushko, Evgeni Mitskevich and Athanasios Mamalis [2]

[1] Institute for Superhard Materials, Kiev, Ukraine

[2] Project Center for Nanotechnology and Advanced Engineering, NCSR "Demokritos", Athens, Greece

ABSTRACT

The results obtained by direct nano-patterning demonstrate the potential of the SPM-based techniques that include surface scratching to create 3D nanostructures. Such techniques became known as tribo-nanolithography and have prospects of being successfully implemented in the future nanofabrication industry. An important obstacle to this, however, is the effect of wear at the nanometer scale which is critical to the stability of tribo-nanolithoraphic processes. Such stability is achievable via in-depth theoretical and experimental studies of friction at the nanoscale along with the development of pioneering equipment. Our work presents the results of experimental fabrication of nanostructures formed by nanoscratching with the use of the multifunctional scanning tunneling microscopy previously developed by the authors. The authors attempted scratching the silicon surface by using a boron-doped diamond tip. This operation was undertaken in the same direction sequentially with the tip sliding a side of the groove by one of the tip's facets and the consequent surface scanning. Although not being applicable to non-conductive surfaces, the proposed technique has significant advantages. One advantage is related to the high stiffness of the tunneling probe as compared to the stiffness of the AFM cantilever. High stiffness and perpendicularity of the tip to the surface during surface processing eliminates bending beam effects on the typical AFM and ensures machining effectiveness. Purposely synthesized boron-doped single-crystal diamonds were used as a tip material. The results of experimental fabrication of nanostructures formed by nanoscratching with the use of the multifunctional scanning probe are demonstrated and discussed.

INTRODUCTION

Normally SPM involves scanning the surface of the sample with minimal (if any) effect on it, and, as a result, forming an image, either of the surface profile or of the distribution of certain physical parameters over the surface. However, the SPM is also able to "actively" affect the surface with the probe and change its characteristics. One such method is nanolithography, in which certain structures with nanometer or even atomic size can be formed on the surface [1-5]. Usually such structures are placed on-surface and, therefore are mechanically and/or thermally sensitive and often require special environmental conditions. SPM-fabricated in-depth structures are more stable and protected. These structures can be obtained by the tribo-nanolithography method [7-9], which involves nanomechanical surface processing. This method of "active" on-surface operations uses SPM as a machine for nanomechanical surface modification. In [10-14] nanomechanical processing methods like cutting hard materials such as Silicon, Gallium, Arsenic compositions, and some metallic surfaces are demonstrated. These studies utilize regular metallic probes as tools as they were initially intended to be used for the surface scanning only. Because of their low wear resistance these soft probes are impractical in surface processing. Instead, utilization of diamond for this purpose can resolve issues associated with wear or breakage of the probe's tip during nanomechanical surface processing. Related works [15-17] show various approaches in using a CVD diamond tip in AFM.

AFM itself, however, involves bending of the cantilever and thus, low rigidity of the tool and its deviation from normality to the surface. Obviously these side effects are highly undesirable. Utilization of the SPM approach with rigid probe is more preferable. Recently, some results of the utilization of SPM equipped with boron-doped tips for complex studies of the surface employing indentation and scanning were reported [18,19]. The present work demonstrates an application of such an approach towards the various nano-tribolithography techniques.

EXPERIMENTAL DETAILS

The material used for the STM tip is boron-doped type IIb diamond single crystals, synthesized by the temperature gradient method under high pressure—high temperature (HPHT) conditions. Studies of the internal structure of crystals with different characteristics have shown that crystals with octahedral structure possess the best required properties, even when octahedral tips are dulled with {100} facets which form minor <100> growth pyramids. Crystals with preferentially developed octahedral facets were used for the tip production. Such crystals have higher conductivity because their volume is formed mainly by <111> growth pyramids which absorb higher quantities of boron dopant. Inclusion-free crystals with clearly formed apexes and crystallographic directions, verified with X-ray tests, were chosen to manufacture the tip with the shape of a pyramidal indenter [20]. A special cast-iron table subjected to the so-called "multi-year ageing" was used as the grinding tool. The working surface of the table has been impregnated with diamond micro-powder with granularity less than 1 mm. The diamond shaping process formed the tip of a Berkovich pyramid with a radius of curvature of about 80 nm as recommended by ISO-14577. The preparation procedure of the probe is described in our previous work [19].

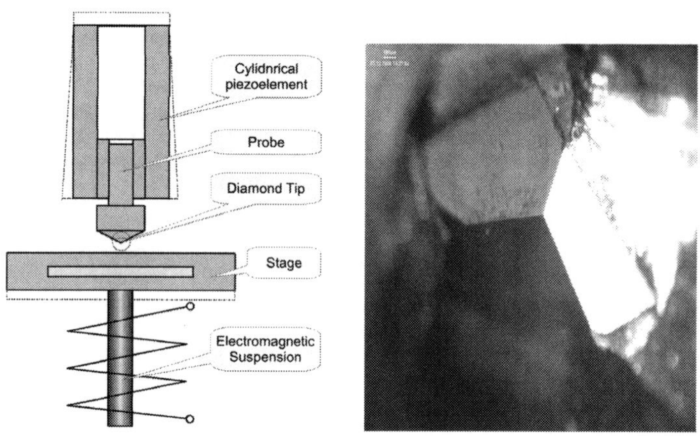

a *b*

Figure 1. Multifunctional SPM with diamond tip: *a* – principal layout of SPM equipped with diamond tip; *b* – diamond tip shaped as Berkovich pyramid.

We use a home-built multifunctional SPM that combines an STM, having a traditional piezo-head, and an electromagnetic loading measurement mechanism, operating while the tip is in contact with the surface. The load is determined via the measurement of the current required to maintain the stage in place. The GUI-based software allows interactive operation of the instrument and performs data processing visualization and storage. Experimental application of the SPM discussed that is based upon STM working principle is limited to the conductive surfaces. Nevertheless, as compared to the regular AFM, this instrument has several substantial advantages. High stiffness of the probe and its perpendicularity to the specimen's surface are the principal ones. This makes possible to eliminate bending beam effect present on the typical AFM and enhance machining performance in comparison with the regular AFM.

RESULTS AND DISCUSSION

Mechanical surface nanoprocessing is the key component of nano-tribolithography. The pressing force between probe and surface is controllable during SPM operation. This makes it possible to produce nanometer-deep furrows on many hard materials.

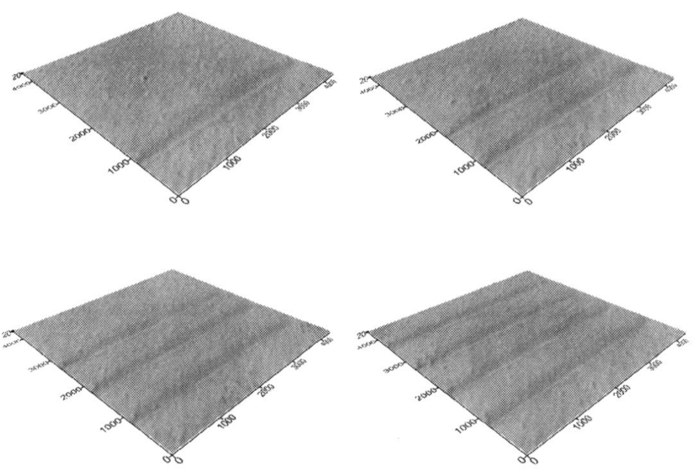

Figure 2. STM image of grooves on the silicon surface made by the diamond pyramidal tip. The groove depth is 12-16 nm.

Figure 2 demonstrates the results of four scratches on the silicon surface by using a boron-doped diamond tip. Scratching has been performed in the same direction. The tip was oriented with regard to the groove such that the probe's axis, an edge adjacent to the tip, and the bottom line of the groove were coplanar. Scanning of the surface subjected to scratching was performed after each operation. As seen in Figure 2, the formation of the groove was not accompanied by the visible on-surface cracking that was expected for a ceramic. Perhaps, the phenomenon known as nanoscale ductile cutting [21], when the surface of a brittle material behaves like a plastic at the nanoscale level takes place here. However, the force distribution under SPM probe interaction with the surface differs substantially from that for micro-cutting machines. Further studies of surface SPM-powered nano-processing with the employment of molecular dynamics methods are required.

Similar operation with smaller penetration depth of the tip into the surface can lead to the removal of some material out of the substrate. An example of such operation is shown on Figure 3. 50 nm-thick gold film deposited in the silicon substrate served as an example. The removal of the gold has been performed in the scanning mode with the same penetration depth of the tip into the surface.

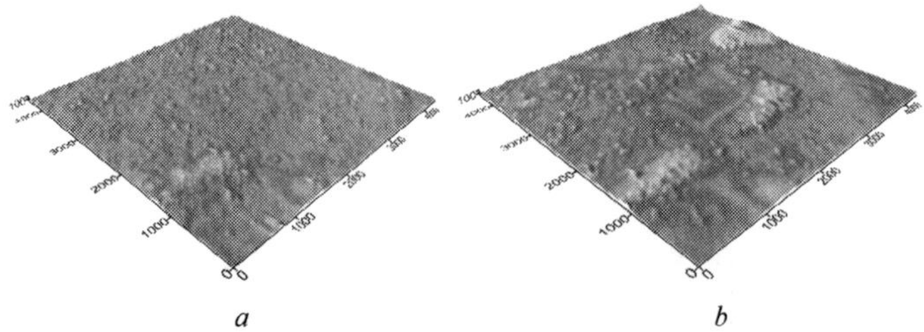

Figure 3. SPM image of the gold film on the silicon substrate after the removal of the single (*a*) and three separate (*b*) rectangular fragments of the gold film.

Experiments have shown high wear resistance of the boron-doped diamond tip while performing nanomechanical processing both on soft (Au) and on hard (Si) surfaces. The tip's radius of curvature was measured by using electron microscopy and after two hundred experiments and remained within about 15 nm. The contamination of the tip and necessity to clean it up periodically has to be mentioned as a drawback of this technique. This is something to be kept in mind while developing industrial-scale applications.

The results shown above allow us to develop technology to produce complicated structures in the depth of the surface.

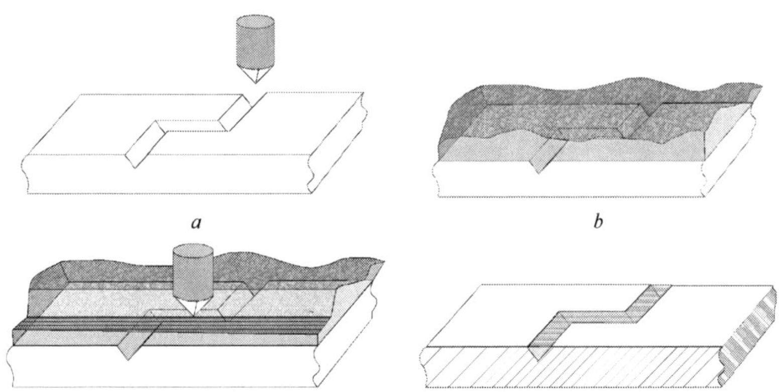

Figure 4. Nano-tribolitographic procedure of the formation of conducting paths on the silicon surface: *a* – initially 10…20 nm-deep paths are scratched out on the silicon surface with the use of monocrystalline diamond SPM probe; *b* - a ~30 nm Au layer is deposited on the silicon surface; *c* – the extra Au is removed by the diamond tip down to the ~10 nm; *d* –the rest of the gold (except from the paths) is chemically etched.

In our technique we use the following approach to form the on-surface structures. During the first stage, 10 to 20 nm-deep paths of the silicon surface are carved by using the SPM probe equipped with the diamond tip. In the next stage, a 30 nm-thick gold layer is deposited on the mechanically modified surface. Later, using the same diamond tip, surface smoothing is performed to leave the gold layer about 10 nm thick. During the last, fourth stage, gold is removed by chemical etching. The obtained structure is more stable from chemical and mechanical influence than that obtained by traditional nano-lithography.

CONCLUSIONS

Experiments on surface nano-processing has demonstrated the effectiveness of the utilization of a single crystal diamond as the material for the probe's tip fabrication. So far no experiments resulted in tip deformation caused by fracture of wear. The utilization of the "cantileverless" SPM technique is preferred over the AFM-based techniques as the STM-probe possesses far higher rigidity and the resolution for surface scanning is independent of the tip's radius of curvature. The presented nano-lithography techniques require further development and will find industrial implementation when combined with multiple SPM-probe systems.

ACKNOWLEDGMENTS

The authors would like to thank Dr. S. Ivachnenko for the synthesis of conductive diamond. This research is supported by the Ukrainian government program "Nanotechnology and Nanomaterials" under contract No. 2.2.11.31.

REFERENCES

1. D. M. Eigler, E. K. Schweizer, *Nature.* **344**, 524 (1990).
2. L. A. Nagahara, T. Thundat, and S. M. Lindsay, *Appl. Phys. Lett.* **57**, 270 (1990).
3. A. Majumdar, P. I. Oden, J. P. Carrejo et al., *Appl. Phys. Lett.* **61**, 2293 (1992).
4. P. Davidsson, A. Lindell, T. Makela et al., *Microelectron. Eng.* **45**, 1 (1999).
5. M. Ishibashi, S. Heike, H. Kajiyama et al., *Appl. Phys. Lett.* **72**, 1581 (1998).
6. J. K. Gimzewski, R. Möller, *Phys. Rev. B.* **36**, 1284 (1987).
7. N.Kawasegi, et al., *Nanotechnology* **16**, 1411 (2005).
8. J.W. Park, N. Morita, and D.W.Lee, *Journal of Nanoscience and Nanotechnology*, **10**, 4440 (2010)
9. J.W. Park, D.W.Lee, N. Takano and N. Morita, *Materials Science Forum*, **79**, 505 (2006).
10. M.A. McCord, R. F. Pease, *Appl. Phys. Lett.* **50**, 569 (1987).
11. E. J. Van Loenen, D. Dijkkamp, A. J. Hoeven et al., *J.Vac. Sci.Technol. A.* **8**, 574 (1989).
12. M. Versen, B. Klehn, U. Kunze et al., *Ultramicroscope.* **82**, 159 (2000).
13. S. Miyake, *Appl. Phys. Lett.* **67**, 2925 (1995).
14. T. H. Fang, W. J. Chang, *J. Phys. Chem. Solids.* **64**, 913 (2003).
15. E. Oesterschulze, A. Malave, U. F. Keyser et al., *Diam. Rel. Mater.* **11**, 667 (2002).
16. K. Ashida, Morita N., Yushida Y., *JSME Int. J., Ser.C.* **44**, 244 (2001).
17. N. Kawasegi, N. Takano, D. Oka et al., *J. Manuf. Sci.Eng.* **128**, 723 (2006).
18. O. Lysenko, N. Novikov, A. Gontar, et al., *J. Phys.: Conf. Ser.* **61**, 740 (2007).
19. O. Lysenko, N. Novikov, V. Grushko et al., *Diam. Rel. Mater.* **17**, 1316 (2008).
20. N. Novikov, T. Nachalna, S. Ivakhnenko et al., *Diam. Rel. Mater.* **12**, 1990 (2003).
21. T.P. Leung, W.B. Lee and X.M. Lu, *J. Mater. Proc. Tech.*, **73**, 42 (1998).

Mater. Res. Soc. Symp. Proc. Vol. 1318 © 2011 Materials Research Society
DOI: 10.1557/opl.2011.163

Photoinduced Temporal Change of Surface-Potential Undulation on Alq3 Thin Films Observed by Kelvin Probe Force Microscopy

Kazunari Ozasa, Hiromi Ito, Mizuo Maeda, and Masahiko Hara
Advanced Science Institute, RIKEN, 2-1 Hirosawa, Wako, Saitama 351-0198, Japan

ABSTRACT

The surface potential (SP) undulation on the surfaces of tris(8-hydroxyquinolinato) aluminum (III) (Alq3) films has been investigated with Kelvin probe force microscopy (KFM) and scanning near-field optical microscope (SNOM)-KFM. The SP undulation observed on the amorphous Alq3 films with thicknesses of up to 300 nm showed a cloud-like morphology of 200–300 nm in lateral size. The temporal change of SP undulation was traced through cyclic measurement with KFM observation with intermittent photoexposure, as well as in situ localized photoexcitation with SNOM-KFM. We concluded that the origin of the SP undulation is the nonuniform distribution of charged traps and drift mobility in the Alq3 films.

INTRODUCTION

Tris(8-hydroxyquinolinato) aluminum (III) (Alq3) is one of the major organic semiconductors for light emitting devices. When an Alq3 film is prepared by vacuum deposition with light shielding, surface potential (SP) proportional to the film thickness is spontaneously formed [1]. Photoexposure reduces the SP of Alq3 [1-4]. We have previously reported the dependence and mechanism of the photoinduced SP reduction, and concluded that the mechanism is the drift of photoexcited carriers according to electric fields in the films and subsequent capturing the carriers by deep traps [3,4].

Recently, we found that the spatial undulation of the SP grows once and then decreases with photoexposure. We traced the temporal change of SP undulation through cyclic measurement with Kelvin probe force microscopy (KFM) with intermittent photoexposure, as well as in situ localized photoexcitation with a scanning near field optical microscope (SNOM)-KFM [5].

EXPERIMENT

Amorphous Alq3 films with various thicknesses up to 300 nm were formed on n-Si(001) by vacuum deposition at room temperature with light shielding at a deposition rate of 0.15-0.2 nm/s. The topographic and SP- morphology were measured with KFM with lift mode.

Intermittent photoexcitation was carried out with a white lamp light for probe positioning in KFM apparatus. Localized photoexcitation was examined with a blue laser through fiber probe of SNOM-KFM [5].

DISCUSSION

For 100-nm thick Alq3 film, which showed an average SP of 4.2 V, the undulation of SP was 13 mV (histogram width) with spatial size 200–400 nm, while the topographic morphology has spatial size 20-30 nm and height variation <2 nm, as shown in Fig. 1. The difference in spatial sizes indicates that the origin of SP undulation is spatial nonuniformity in the dipole alignment of Alq3 molecules, not the film thickness fluctuations.

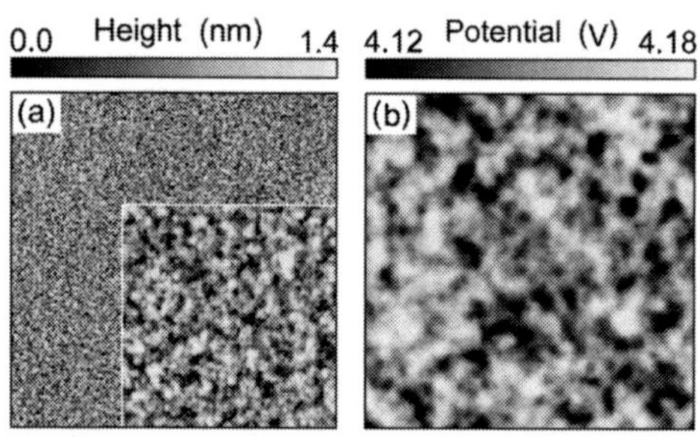

Figure 1. Topographic image (a) and SP image (b) observed on a 100-nm-thick Alq3 film deposited on n-Si(001), with a scan size of 5 μm.

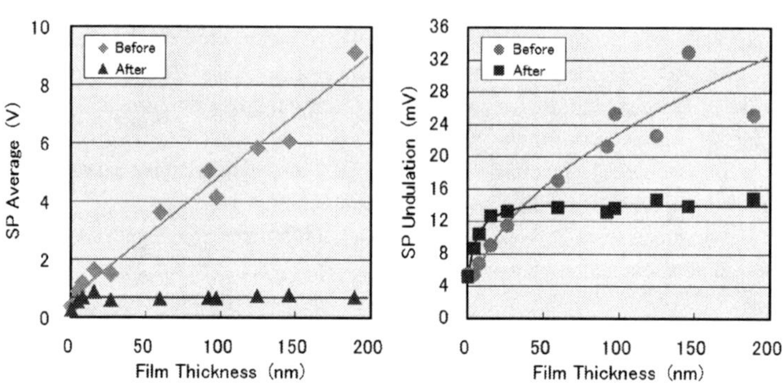

Figure 2. Thickness dependences of SP average and SP undulation, before photoexposure and after a 900-s photoexposure.

The average SP depends linearly on Alq3 film thickness, while the amplitude of SP undulation shows a sublinear dependence on the film thickness, as shown in Fig. 2. The SP undulation in Fig. 2 was calculated as the full width of half maxima (FWHM) of the SP histogram (Gaussian fitting) of the corresponding SP image with a 5-μm scan area, and therefore, was approximately 2.5 times smaller than that of the peak-to-peak amplitude. The curves for SP average and SP undulation before the photoexposure in Fig. 2 are with linear and square root fitting curves, respectively.

The origin of the SP undulation can be either the spatial fluctuation of the alignment of the Alq3 molecular dipole moments or the nonuniform distribution of charged traps. The two possibilities can be examined by investigating the photoexposure effects, since the drift of photoexcited carriers will drastically affect only to the nonuniform distribution of charged traps. When the surface was exposed to a white light, the SP averages of the Alq3 films were all reduced to approximately 0.7 V, while the SP undulations were reduced to a constant level of 14 mV for film thicknesses larger than 25 nm. The photoinduced large change of the SP undulation revealed that the nonuniform distribution of the charged traps, rather than the spatial fluctuation of the alignment of the Alq3 dipole moments, is the origin of the SP undulation. The SP undulation remained after longer photoexposure over 30 min, revealing that the nonuniform distribution of charged traps cannot be canceled out completely. The residual SP undulation was mostly independent of the Alq3 film thickness. This suggests that the distribution nonuniformity of the charged traps is localized at the surface/interface and not in the film bulk.

Figure 3 shows the photoinduced temporal change of the SP average and SP undulation for a 100-nm-thick Alq3 film, obtained by KFM observation and intermittent photoexposures. The line profiles of SP undulation taken sequentially at the same position during the photoexposure are given in the inset in Fig. 3.

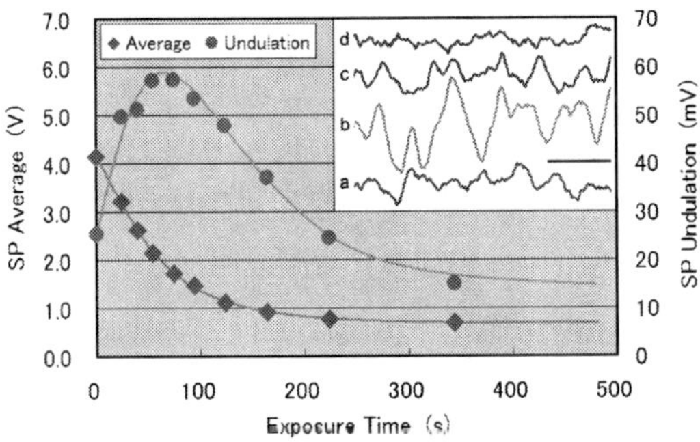

Figure 3. Temporal evolution of SP average and SP undulation during photoexposure, observed for a 100-nm-thick Alq3 film. The inset shows the line profiles of SP undulation at the photoexposure of (a) 0 s, (b) 74 s, (c) 164 s, and (d) 344 s, obtained at the same observation line. Scale bar in the inset shows 1μm.

Remarkably, a short photoexposure for 0–74 s increased the SP undulation with decreasing the SP average, where several new and large peaks were observed on the line profile. Further photoexposure reduced the SP undulation, and the peaks appeared on the line profile faded out. The period of the enhancement of the SP undulation corresponds to that of the rapid decrease in the SP average. The correspondence indicates that the drift of the photoexcited carriers enlarges the SP undulation, i.e., the nonuniform distribution of charged traps. We consider that the dynamic movement of the photoexcited carriers is not uniform in the films due to the nonuniformity of drift mobility in the Alq3 films. The drift mobility is closely related to the trap distribution via trap-filled conduction [6-8]. Therefore, the nonuniform distribution of the charged traps leads to the nonuniformity of drift mobility in thin Alq3 films.

The effects of localized photoexcitation on the SP undulation were examined with SNOM-KFM, where a fiber probe was repeatedly line-scanned for a 10-μm distance with photo illumination from its apex. After certain duration of the line-illumination, the SP morphology was observed with the same fiber probe. As localized photoexcitation proceeded, a line drop of SP was observed as shown in Fig. 4. The SP drop was limited within 3–4 μm, and the SP undulation pattern out of the SP-drop region remained. This indicates a spatial limit in the effects of photoexcited carriers in Alq3 thin films.

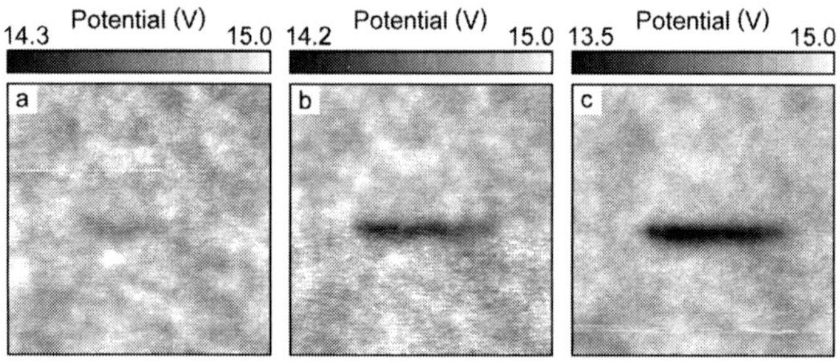

Figure 4. The SP image observed by localized photoexcitation through the apex of a fiber probe of SNOM-KFM with repeatedly scanning the probe on a fixed 10-μm line, for (a) 4 min, (b) 12 min, and (d) 32 min. Scan size was 20 μm.

Other research groups have also reported the localized charged traps at the interfaces, which affect to the luminescence intensity and the lifetime of Alq3 organic light emitting devices (OLEDs) [9-12]. The results in this study show that the nonuniform distribution of charged traps in thin Alq3 films (and probably in other organic semiconductor films) will severely affect on the drift mobility in the films, which will also influence the device performances of OLEDs. Further investigation on the SP undulation especially that with hetero interfaces will be required to figure out the spatial distribution of interface charge traps.

CONCLUSIONS

The SP undulation observed on the thin Alq3 films shows a cloud-like morphology of 200–300 nm in lateral size, which has been attributed to the nonuniform distribution of charged traps in the Alq3 films. The photoinduced dynamic behaviors of the SP undulation were observed with KFM and SNOM-KFM, indicating the nonuniformity of drift mobility in the Alq3 films. Further investigation on the SP undulation observable with KFM or SNOM-KFM will give useful information on the dipole moment distribution, carrier traps near the surface, and the dynamics of photoexcited carriers in thin organic semiconductor films other than Alq3 as well.

ACKNOWLEDGMENTS

The authors thank Dr. Takashi Isoshima and Dr. Eisuke Ito at RIKEN for their suggestive comments. The authors acknowledge financial support for this study by the Ministry of Education, Science, Sports and Culture, Grant-in-Aid for Scientific Research (B), Grant No. 16360022, 2004–2006.

REFERENCES

1. E. Ito, Y. Washizu, N. Hayashi, H. Ishii, N. Matsuie, K. Tsuboi, Y. Ouchi, Y. Harima, K. Yamashita, and K. Seki, J. Appl. Phys., 92 (2002) 7306.
2. K. Sugi, H. Ishii, Y. Kimura, M. Niwano, E. Ito, Y. Washizu, N. Hayashi, Y. Ouchi, and K. Seki, Thin Solid Films 464 (2004) 412.
3. K. Ozasa, S. Nemoto, T. Isoshima, E. Ito, M. Maeda, and M. Hara, Surf. Interface Anal., 40 (2008) 579.
4. K. Ozasa, S. Nemoto, T. Isoshima, E. Ito, M. Maeda, and M. Hara, Appl. Phys. Lett., 93 (2008) 263304.
5. K. Ozasa, S. Nemoto, M. Maeda, M. Hara, J. Appl. Phys., 107 (2010) 103501.
6. P. E. Burrows, Z. Shen, V. Bulovic, D. M. McCarty, S. R. Forrest, J. A. Cronin, M. E. Thompson, J. Appl. Phys., 79 (1996) 7991.
7. S. Karg, J. Steiger, and H. von Seggern, Synthetic Metals, 111-112 (2000) 277.
8. A. G. Werner, J. Blochwitz, M. Pfeiffer, and K. Leo, J. Appl. Phys., 90 (2001) 123.
9. T. Mori, S. Miyake, and T. Mizutani, Jpn. J. Appl. Phys., 34 (1995) 4120.
10. H. Fong, K. Lun, and S. So, Jpn. J. Appl. Phys., 41 (2002) L1122.
11. Y. Nakano, K. Noda, H. Fujikawa, and T. Morikawa, Jpn. J. Appl. Phys., 47 (2008) 464.
12. Y. Noguchi, N. Sato, Y. Miyazaki, and H. Ishii, Appl.Phys.Lett., 96 (2010) 143305.

Mater. Res. Soc. Symp. Proc. Vol. 1318 © 2011 Materials Research Society
DOI: 10.1557/opl.2011.140

Atomic force microscopy based quantitative mapping of elastic moduli in phase separated polyurethanes and silica reinforced rubbers across the length scales

Peter Schön[1], Kristóf Bagdi[2,3], Kinga Molnár[2,3], Patrick Markus[4], Saurabh Dutta[1], Morteza Shirazi[5], Jacques Noordermeer[5], Béla Pukánszky[2,3] and G. Julius Vancso[*,1]

[1]Materials Science and Technology of Polymers, MESA+ Institute for Nanotechnology, University of Twente, Enschede, NL-7500, The Netherlands
[2]Laboratory of Plastics and Rubber Technology, Department of Physical Chemistry and Materials Science, Budapest University of Technology and Economics, P.O. Box 91, H-1521 Budapest, Hungary
[3]Institute of Materials and Environmental Chemistry, Chemical Research Center, Hungarian Academy of Sciences, P.O. Box 17, H-1525 Budapest, Hungary
[4]Bruker Nederland B.V., Bruynvisweg 16, 1531 AZ Wormer, The Netherlands
[5] Elastomer Technology and Engineering, Engineering Technology Department, University of Twente, Enschede, 7500 AE, the Netherlands

*Corresponding author:
Prof. G. Julius Vancso, Tel: +31-53-4892967
Fax: +31 (0)53 489 3823
E-mail: g.j.vancso@utwente.nl

ABSTRACT

In the work presented here atomic force microscopy (AFM) based mechanical mapping techniques - HarmoniX imaging and Peak Force Tapping - were applied to determine the surface elastic modulus of phase separated polyurethanes and silica reinforced rubbers across the length scales. Segmented polyether polyurethanes (PUs) were prepared with varying stoichiometric ratio of the isocyanate and hydroxyl groups. The effect of molar mass, as well as the type and number of end-groups on their morphology was investigated. Smooth PU samples for AFM imaging were prepared by ultramicrotonomy. The micro phase separated morphology of the phase separated PUs showed characteristic "fingerprint" AFM phase images. Surface modulus values obtained by AFM were compared to bulk modulus values obtained by tensile testing. The moduli were mapped quantitatively with nanoscale resolution and were in excellent agreement for both AFM modes. Surface mean moduli values do not coincide with bulk values obtained via tensile testing which is attributed to fundamentally different averaging procedures and effects that lead to the respective modulus values obtained via surface and volume averaging. EPDM and SBR rubbers and rubber blends thereof were prepared with varying concentrations of silica nanoparticles and studied in order to investigate the effect of different composition on the resulting morphology (filler distribution) and elastic moduli on a specific rubber or rubber blend sample. Elastic moduli of the rubber and rubber blend samples were first measured by bulk tensile testing. The morphology of the rubber samples was visualized by height and phase imaging. Surface elastic moduli of silica reinforced rubbers and rubber blends were mapped quantitatively and compared with bulk tensile test results. AFM allowed the determination of modulus distributions at the sections imaged. As potential reasons for the observed differences between bulk and surface modulus different averaging procedures like surface and bulk averaging of AFM vs. tensile testing, different filler distributions in SBR and EPDM and the AFM modulus calibration procedures can be named.

INTRODUCTION

Polyurethanes represent a versatile class of soft matter with a rich variety of applications. According to the intended end-use with the proper control of their molecular structure PUs can be designed to possess either soft or stiff mechanical characteristics [1, 2]. The composition of the phases is a priori unknown, although the amount of soft and hard segments can in principle be determined from the composition of the reaction mixture. In addition, segmented polyurethanes were shown to form hierarchical structures with several structural units at different length scales [3]. For a complete elucidation of the PU morphology the effect of these structural units on the bulk properties and the dominating structural element must be characterized.

Rubbers and rubber blends represent another class of highly important materials due to their unique mechanical properties. For practical applications, rubber compounds contain a significant amount of various ingredients, amongst which silica nanoparticles are used as reinforcing filler for mechanical strengthening, for example [4]. The homogeneity of the filler distribution is a fundamentally important parameter influencing the properties of rubber blends. However the use of this technique is limited for the detection and localization of SiO_2 nanoparticles. Transmission electron microscopy requires sophisticated preparation of microtome section. The dispersion of fillers in rubbers has been determined predominantly by optical and transmission electron microscopy (TEM) until recently.

AFM has been widely used to study the structure and morphology of polymers [5, 6]. Microphase separated structures of polyurethanes were investigated by AFM tapping mode height and phase imaging [3]. The rubber morphology and carbon black microdispersions have also been studied by atomic force microscopy [7, 8]. Importantly, quantitative mechanical mapping was not possible with conventional phase imaging as the phase signal is related to the energy dissipation of the tapping tip which cannot be directly translated into the variation of mechanical moduli [9]. Quantitative mechanical information on surface moduli at polymer surfaces was obtained by nanoindentation and force volume AFM imaging [10-12]. Local elastic moduli can be derived from the recorded force vs. deformation curves using appropriate mechanical contact models. However, serious drawbacks for nanoindentation and force volume imaging are the poor lateral resolution and the extensive time required to obtain a complete mapping of the surface.

Recently AFM based HarmoniX Imaging and Peak Force Tapping have been introduced as mechanical property mapping techniques [13]. Both techniques provide high resolution (< 5 nm laterally) quantitative mechanical characterization of stiffness, elastic modulus, adhesion and energy dissipation of a polymer surface. In the work presented here HarmoniX Imaging and Peak Force Tapping were applied to map the surface elastic moduli of phase separated polyurethanes and silica reinforced rubbers across the length scales.

EXPERIMENT

Preparation of polyurethane samples

Polyurethanes were prepared from 4,4'-methylenebis(phenyl isocyanate) (MDI), 1,4-butanediol chain extender (BD) and poly(tetrahydrofurane) polyether polyol (PTHF) with a molar mass of 1000 g/mol under the conditions described earlier. After tensile testing of the specimen (cross-section about 2 mm x 2 mm) the respective PU sample was frozen in liquid nitrogen and cut by a glass knife using a cryomicrotome (Leica Ultracut, Leica, Wetzlar, Germany) for AFM measurements. After cryomicrotoming the samples were glued to a metal support using a two component epoxy and dried overnight.

Preparation of rubber compounds

EPDM and SBR compounds were prepared under variation of the amount of silica filler. The amount of silane (Si 69®) and paraffinic oil (Sunpar 150) needed to functionalize and disperse the silica particles was also varied, depending on the amount of silica. The concentration of all other components like ZnO (activator), stearic acid (co-activator), TMQ (antioxidant) and curatives was kept the same, for each series of SBR and EPDM compounds. Additional blend samples were made with different ratios of EPDM/SBR with the same amount of silica (20 phr) to study the distribution of silica in the two phases. For AFM imaging a piece of the cured rubber sample was cut using a scissor followed by cutting a small piece from the strip using a sharp razor blade so as to obtain a smooth surface and avoid any rough edges. The sample was glued to a metal disc with a two -component epoxy glue. The glue was allowed to dry overnight. The sample surface was cleaned with nitrogen before AFM imaging.

Atomic Force Microscopy

Torsional harmonic cantilevers (HarmoniX Probes, HMX, HMX-S, Veeco Instruments) (resonance frequency ~60 kHz, spring constant 1-1.8 N/m) were used for AFM based HarmoniX mapping. Measurements were done in air under ambient conditions using a Multimode AFM with a NanoScope V controller and NanoScope version 7.30 software (Veeco/Digital Instruments, Santa Barbara, CA). For quantitative DMT modulus measurements the cantilevers were calibrated using a standard PS/LDPE sample. HarmoniX Tapping Mode AFM was operated with a cantilever vibration free amplitude of 1.5 V in air. The force applied to the surface was adjusted by the amplitude setpoint, which was used for the feedback control, to ca. 40% of the free amplitude. Imaging was done at 0.5-1 Hz scan rates. Peak Force Tapping was performed with Si tips on SiN cantilevers (SCANASYST-AIR, Veeco Instruments Probes, Camarillo, CA) under ambient conditions in air. A Multimode 8 AFM equipped with a NanoScope V controller and NanoScope version 8.10 software (Veeco/Digital Instruments, Santa Barbara, CA) was used. Data analysis and image processing and were performed with the NanoScope software version 8.10, and NanoScope Analysis software version 1.10.

DISCUSSION

Polyurethane samples

Three linear PU elastomers with varying ratios of the isocyanate and hydroxyl groups were used for AFM studies. The –OH functional group ratio of polyol/total diol (POLY(OH)/OH ratio) was kept constant at 0.4 in all experiments. The variable was the ratio of the isocyanate and hydroxyl groups (NCO/OH), which changed between 0.940 and 1.150 in 16 steps for the series. The composition and specimen identification codes of the three samples selected from the series for AFM studies are listed in Table 1.

Table 1 Composition of the PU elastomers used for AFM studies

Label	NCO/OH	MDI	PTHF	BD
		Mmol	mmol	mmol
2/01	0,940	69,48	29,57	44,35
2/09	1,025	73,76	28,78	43,18
2/16	1,150	79,66	27,71	41,56

PU topography shows 'fingerprint' phase images

Freshly cryomicrotomed PU samples were investigated by AFM HarmoniX Imaging. Phase images were recorded for 2µm x 2µm scan sizes for each sample (Fig. 1). The hard segment motifs show similar widths of ca. 20 nm in all three samples. However, the global features of the morphologies vary significantly. AFM phase imaging was highly reproducible revealing the characteristic features of the specific morphologies described on several cryomicrotomed specimens for each sample. It was possible to identify each sample by the appearance of its phase morphology, which in this way represent a "fingerprint" image.

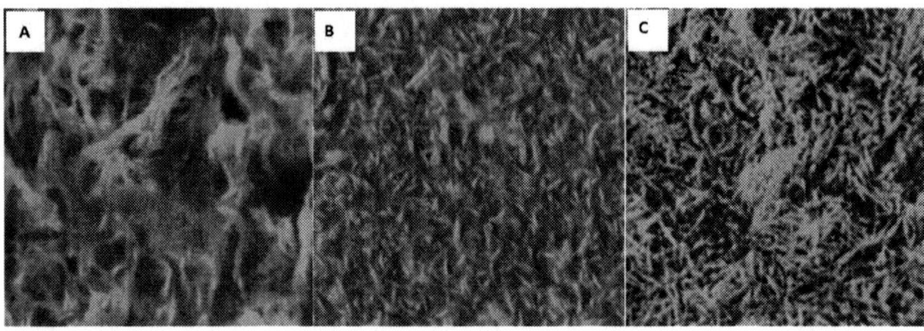

Figure 1. Phase images as obtained by HarmoniX Tapping mode AFM. A: sample 2/01; B: sample 2/09 and C: sample 2/16; scan sizes: 2 µm x 2 µm. *z*-scale: 30°.

DMT moduli maps have been recorded with AFM based HarmoniX imaging. DMT moduli maps (Fig. 2) reveal significantly differently morphologies, as was observed already in the corresponding phase images. Measurements were complemented with Peak Force Tapping. Again similar morphologies in the DMT modulus mappings were visualized. As an example the modulus map of sample 2/16 is shown in Figure 2 including the corresponding modulus distribution. Both AFM modes clearly resolve elastic moduli of soft and hard segments of the phase separated polyurethane samples with nanoscale resolution.

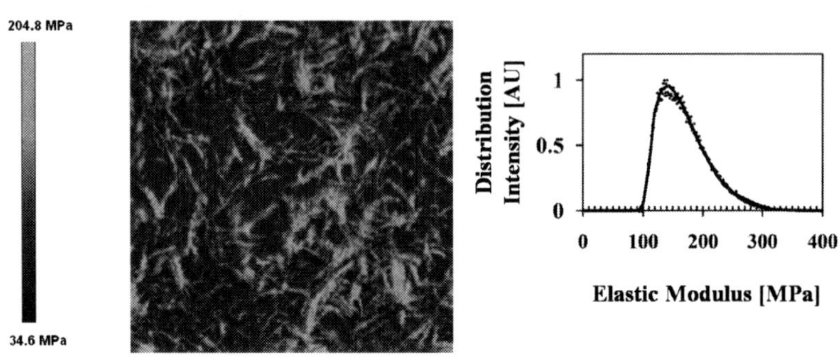

Figure 2. Example of a DMT Modulus map and distribution as obtained by Peak Force Tapping; sample 2/16; scan size: 2 µm x 2 µm.

Silica reinforced rubber samples

DMT modulus maps obtained from silica reinforced SBR are shown in Figure 3. Isolated particle morphologies can be observed with diameters ranging from 20-80 nm; the corresponding DMT modulus values range from 0.8 to 2 GPa. The surrounding matrix reveals moduli between 200 – 300 MPa. Comparing the images of SBR-5 phr silica, SBR-10 phr silica and SBR-20 phr silica, an increase in the number of particles per area with higher modulus can be observed. Since the increase in the number of stiffer particles corresponds to the increase in the concentration of silica in the compound preparation, and the sizes observed are similar to those of primary silica particles, the particles are identified as the silica filler. In SBR, the Ultrasil 7005 silica was predominantly present as well dispersed primary particles and smaller aggregates.

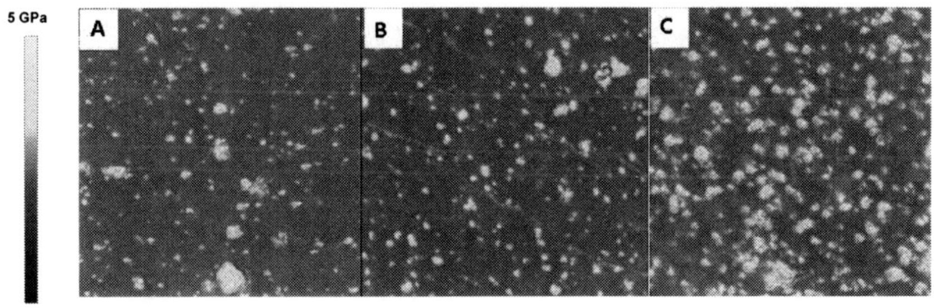

Figure 3. DMT modulus (A,B,C) images as obtained by HarmoniX Tapping mode AFM. Samples: SBR filled with increasing concentration of Ultrasil 7005; A: 5 phr Ultrasil 7005; B: 10 phr Ultrasil 7005 and C: 20 phr Ultrasil 7005; scan sizes: 5 µm x 5 µm. z-scale: 0-5 GPa.

HarmoniX AFM measurements revealed the formation of larger silica aggregates in EPDM in contrast to SBR where isolated silica particles can be observed. At silica reinforced blends of EPDM/SBR a phase separation into a soft (ca. 40 MPa) and a significantly harder phase could be observed (ca. 500 MPa- 1.5 GPa) indicating the incorporation of silica particles exclusively in the SBR phase. In case of SBR/EPDM blends, a phase separation was observed and the two phases could be distinguished by the large differences in their stiffness. While the modulus values in the stiffer phase of the EPDM/SBR were observed to be as high as 2-3 GPa, in the soft part the values were as low as 30-40 MPa.

CONCLUSIONS

The morphology observed at the different PU samples strongly depends on the ratio of the isocyanate and hydroxyl groups (NCO/OH) resulting in specific "fingerprint" phase images which allow the identification of the sample. AFM resolves the elastic moduli of stiff and soft segments of the phase separated polyurethane samples at near-nanoscale resolution. The DMT modulus values recorded with HarmoniX Imaging and Peak Force Tapping are in very good agreement. Surface mean moduli values do not coincide with bulk values obtained via tensile testing which is attributed to fundamentally different averaging procedures and effects that lead to the respective modulus values obtained via surface and volume averaging.

The DMT moduli maps of silica reinforced rubbers and rubber blends obtained by HarmoniX AFM imaging revealed an increase of the areal fraction of silica particles with rising concentration in the compound preparation mixture. Significantly higher elastic moduli were observed by AFM then by bulk tensile testing. While the stiff silica particles show modulus values as high as 2 GPa, the rubber matrix reveals modulus values in the range of 30 MPa (in case of rubber blends) to 300 MPa (individual rubbers). As potential reasons for the observed differences between bulk and surface modulus different averaging procedures like surface and bulk averaging of AFM vs. tensile testing, different filler distributions in SBR and EPDM and the AFM modulus calibration procedures can be named.

ACKNOWLEDGMENTS

P.S. and G.J.V. thank the MESA+ Institute for Nanotechnology for funding. G.J.V. acknowledges support by Veeco in form of the Veeco HarmoniX research grant.
The research on segmented PU was partly financed by the National Scientific Research Fund of Hungary (OTKA Grant No. K 68748), the support is highly appreciated.
M.S. and J.N. acknowledge support from the Dutch Polymer Institute (DPI) under project #664.

REFERENCES

1. Pukánszky, B.; Bagdi, K.; Tovolgyi, Z.; Varga, J.; Botz, L.; Hudak, S.; Doczi, T.; Pukánszky, B. *Eur Polym J* 2008, *44*, 2431.
2. Bagdi, K.; Molnár, K.; Pukánszky, B.; Pukánszky, B. *J Therm Anal Calorim* 2009, *98*, 825.
3. Tocha, E.; Janik, H.; Debowski, M.; Vancso, G. J. *J Macromol Sci Phys* 2002, *B41*, 1291.
4. W. K. Dierkes, J. W.M. Noordermeer. "Silica-Filled Rubber Compounds" in *RubberTechnologist's Handbook*, Volume 2; J.White, S.K. De and K. Naskar, Eds.; Smithers-Rapra, Shawbury UK, 2009, 97-158.
5. Vancso GJ, Schön, P., Duvigneau, J. What's New in Atomic Force Microscopy of Polymers? An Update Microscopy and Analysis 2009;23:5-11.
6. Vancso GJ, and Schönherr, Holger. Scanning Force Microscopy of Polymers: Springer; 2010.
7. van Bevervoorde-Meilof EWE, van Haeringen-Trifonova D, Vancso GJ, Van der Does L, Bantjes A, Noordermeer JWM. *Kaut. Gummi Kunstst.* 2000;*53(7-8)*:426.
8. Trifonova-Van Haeringen D, Schönherr H, Vancso GJ, van der Does L, Noordermeer JWM, Janssen PJP. *Rubber Chem Technol.* 1999;72(5):862-875.
9. Garcia R, Perez R. Dynamic atomic force microscopy methods. Surf Sci Rep. 2002;47(6-8):197-301.
10. Tranchida, D.; Piccarolo, S.; Soliman, M. *Macromolecules* **2006**, *39*, 4547.
11. Schönherr, H.; Feng, C. L.; Tomczak, N.; Vancso, G. J. *Macromol Symp* **2005**, *230*, 149.
12. Vancso, G. J., and Schönherr, Holger *Scanning Force Microscopy of Polymers*; Springer, 2010.
13. (a) Sahin, O.; Magonov, S.; Su, C.; Quate, C. F.; Solgaard, O. *Nat Nanotechnol* 2007, *2*, 507. (b) Sahin, O. *Rev Sci Instrum* 2007, *78*. (c)Sahin, O.; Erina, N. *Nanotechnology* 2008, *19*.

Mater. Res. Soc. Symp. Proc. Vol. 1318 © 2011 Materials Research Society
DOI: 10.1557/opl.2011.141

In Situ Chemical Functionalization of a Single Carbon Nanotube Functionalized AFM Tip using a Correlated Optical and Atomic Force Microscope

Ifat Kaplan-Ashiri, Eric J. Titus and Katherine A. Willets

Department of Chemistry and Biochemistry, The University of Texas at Austin, Austin, Texas, 78712

ABSTRACT

We present a method for performing nanoscale wet chemistry on single carbon nanotubes as well as spectroscopic characterization of the functionalized molecules using a coupled atomic force microscope (AFM) and optical microscope. An AFM probe was functionalized with a single multiwalled carbon nanotube and then locally oxidized by dipping it into nitric acid (HNO_3) *in situ* using AFM manipulation. Raman scattering was collected from the carbon nanotube functionalized probe before and after the oxidation reaction. An increase in the Raman D band was observed after the acid treatment, demonstrating that oxidation had occurred. This is the first step towards developing a real-time technique for dynamic studies of chemical reactions on single nanoparticles/molecules.

INTRODUCTION

Chemical interactions at the nano and molecular scale have been widely studied by scanning probe microscopy (SPM) by means of the interactions between chemically functionalized SPM probes and chemically functionalized surfaces [1-3]. The chemical interactions can be quantified as the adhesive force that is measured between the molecule on the tip and the molecule on the surface; this technique is termed chemical force microscopy [4]. Complimenting chemical force microscopy with optical spectroscopy techniques, such as Raman scattering, can provide us with additional chemical and structural information about nanoscale materials that is unavailable by chemical force measurements alone. For example, the coupling of an atomic force microscope (AFM) with an optical microscope allows real time studies of nanomaterials[5] and single molecules as they undergo chemical changes or interact with the environment around them; this is of great importance for many studies in chemistry and biology.

This paper presents a procedure for performing *in situ* wet chemistry on single carbon nanotube-functionalized AFM tips, which is monitored via Raman spectroscopy. Carbon nanotubes are well-studied in the literature and have a host of interesting mechanical, electronic, and optical signatures that make them potentially useful in a wide array of applications, including field effect transistors [6], light emitting diodes [7], additives to lithium-ion batteries[8] and multifunctional fillers of composite materials [9]. Furthermore, the chemistry of carbon nanotubes has been widely studied, and they can be functionalized with a variety of organic molecules[10]. For instance, carbon nanotubes can be oxidized to produce carboxylic acid groups upon reaction with strong acids like HNO_3, HCl and H_2SO_4[11, 12]. This oxidation reaction serves in many cases as the first step for functionalization of carbon nanotubes with thiol groups[13], DNA[14] and more[10]. Here we demonstrate a technique for site-specific chemical oxidation of a single carbon nanotube using *in-situ* wet chemistry on an AFM platform. By dipping a single carbon nanotube into a small drop of acid and comparing its Raman spectrum before and after reaction, we detect a change in the Raman disorder band which is indicative of chemical oxidation. The approach outlined here is general and can be implemented for the study of many different nano and molecular scale systems. Furthermore, this technique is the first step toward developing real time spectroscopic detection of chemical reactions at the molecular scale level.

EXPERIMENT

In situ wet chemistry and Raman studies of single multiwalled carbon nanotubes were carried out using a coupled AFM and optical imaging/spectroscopy system as shown schematically in Figure 1. AFM tips were functionalized with single multiwalled carbon nanotubes using a scanning electron microscope (SEM); an example of a functionalized AFM tip is shown in Figure 1B. Multiwalled carbon nanotubes, 15 – 20 nm in diameter, were obtained from Xidex Corporation, Austin, Texas and were used as received. Gold-coated silicon AFM probes were purchased from Budget sensors (Tap300GD) and were functionalized with carbon nanotubes as described previously [15]. The nanotubes were mounted on the AFM probes by *in-situ* manipulation in an SEM/FIB (FEI Strata™ DB235 SEM/FIB) equipped with a nanomanipulator (Zyvex s-prober). Each nanotube was welded to the AFM probe by electron-beam deposition of platinum and was checked for whether it fit the length requirements of our experiment (500-1000 nm) before use.

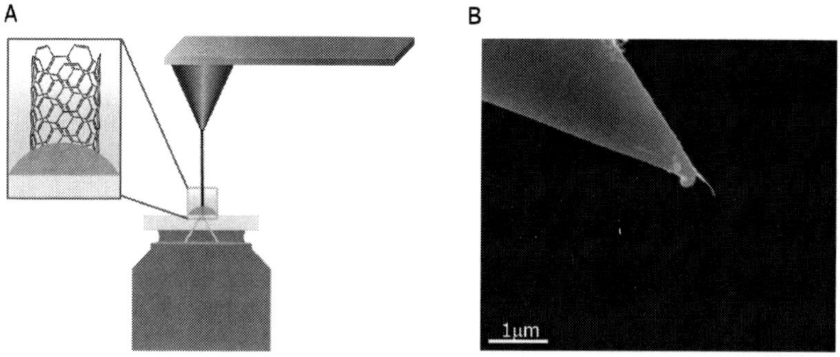

Figure 1: A) Schematic of nanoscale *in-situ* wet chemistry AFM. A single carbon nanotube was dipped into a drop of acid and was detected before and after oxidation with Raman spectroscopy. B) SEM image of the carbon nanotube functionalized AFM probe used in these experiments.

An AFM (NTEGRA Vita, NT-MDT) was mounted on top of an inverted optical microscope (Olympus IX-71) while the objective lens (Olympus 60x, NA = 1.45, total internal reflection fluorescence (TIRF) oil immersion) was mounted at the base of the AFM. Silver colloids were drop cast from solution onto argon-plasma cleaned glass coverslips to serve as focus markers for the experiment. An inorganic immersion liquid (Cargille OHZB) with refractive index of 1.5 was used to provide contact between the objective and the glass coverslip.

White light total internal reflection (TIR) imaging was used to set the focus of the system by imaging the scattering from the silver colloids [5]. The nanotube tip was then mounted in the AFM and allowed to approach the surface. Since we can visually locate the position of the tip on the surface using white light TIR imaging[5], it is possible to position the tip so it will be aligned to the center of the objective (by moving it with the AFM scanner). The nanotube was tapping at the same position during the measurements using the closed loop function of the AFM.

Raman scattering was excited by a 532 nm laser (CrystaLaser CL-2000) introduced via through-the-objective TIR through reflection off a dichroic mirror into the TIRF objective. The scattered light was collected through the same objective, filtered through a notch filter (Semrock), and directed to a spectrometer (PI-Acton Spectra-Pro SP2500), equipped with a 1200 groove/mm grating. The dispersed light was then sent to a liquid nitrogen cooled camera (PI-Acton SPEC-10) for spectral acquisition. The Raman spectrum from a single nanotube was acquired with 10 summed accumulations with 30 second integration times for improved signal-to-noise.

For site-specific chemical oxidation of the carbon nanotube functionalized AFM tip, a 0.5 μL drop of 7% HNO_3 was dropped at the center of the plasma cleaned glass cover slip mounted on the optical microscope and allowed to evaporate. After alignment of the relative positions of the cantilever and the acid drop (see below), the nanotube tip landed at the edge of the evaporated drop and tapped on the surface for one minute. Following oxidation of the nanotube by the acid drop (the CNT tip was not washed with DI water following the oxidation), a new clean glass cover slip was mounted into the microscope and the Raman spectra were acquired again as described above. Additional details are discussed below.

RESULTS and DISCUSSION

Nanoscale wet chemistry *in-situ* AFM can be performed by functionalization of an AFM probe with a desired molecule or nanoparticle, followed by reaction of the functionalized probe with deposited reactant on the sample surface. As described in Figure 1, we have functionalized an AFM probe with a single multiwalled carbon nanotube and oxidized the end of the nanotube by dipping it into a small acid (HNO_3) drop. The relative positions of the acid drop and the cantilever were imaged prior to landing using the AFM optical system with white light brightfield excitation (see Figure 2). We have found that landing directly into the drop can damage the functionalized tip by breaking the nanotube (as the force required to penetrate the surface tension of the drop is too large); as such, we allow the drop to evaporate from the surface before approaching with the nanotube functionalized tip. Although the extent of evaporation is difficult to quantify, the optical images in Figure 2 show a clear difference between the original drop and the evaporated drop, and the lack of damage to the tip upon landing supports our hypothesis that the original drop has largely evaporated.

Using our optical microscope as a guide, we approach a region of the sample surface that was covered by acid prior to evaporation (usually at the edge of the original drop). Since the experiment is done under ambient conditions, the condensation of water vapor at the interface of the nanotube and the glass surface forms a water meniscus, which can serve as the medium for the chemical reaction between HNO_3 remaining on the surface and the carbon nanotube [16]. In addition, the acid concentration can increase locally (due to evaporation), hence a highly reactive environment is formed at the interface of the nanotube and the surface. The result is a highly localized site for chemistry at the nanoscale.

The Raman spectra of the carbon nanotube tip before and after the acid treatment are presented in Figure 3. The Raman spectrum of multiwalled carbon nanotubes has two characteristic peaks, the G band (graphitic band) at ~1580-1590 cm^{-1} and the D band (the disorder band) at ~1350 cm^{-1}. The G band is linked to the graphitic in-plane vibrations (sp^2 hybridization) while the D band is linked to defects in the structure (in the form of sp^3 hybridization, missing atoms in the lattice, bound species, etc) [17, 18].

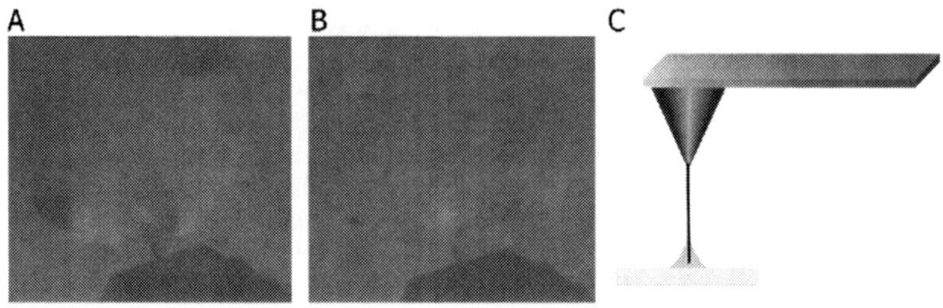

Figure 2: A) Brightfield image of the cantilever and the HNO_3 drop (the tip is away from the surface). B) Brightfield image of the same region after the acid drop is allowed to evaporate. C) A scheme of the water meniscus formed around the nanotube, which serves as a medium for the chemical reaction to occur.

Both the G and the D band Raman peaks are observed in the spectra before and after the acid treatment (Figure 3). However, following acid treatment, the intensity of the D band increased, while the intensity of the G band remained the same. By fitting each spectrum as the sum of two Lorentzians, we can calculate the D/G intensity ratio, which is commonly used as a signature of structural defects in carbon nanotubes induced by chemical modification[19, 20]. Calculating the D/G intensity ratio shows an increase from 0.47 before the acid treatment to 0.59 after the acid treatment. This increase can be attributed to the oxidation of the nanotube and the formation of carboxylic groups on the carbon nanotube. Oxidation of carbon nanotubes by various acids like H_2SO_4, HCl and HNO_3 has been widely studied before [11, 12]. In most of these studies, long reaction times (5-24 hours) were required to oxidize the nanotubes[21], while in our case, we showed that the reaction can occur quite rapidly, as the nanotube was only exposed to the HNO_3 for one minute.

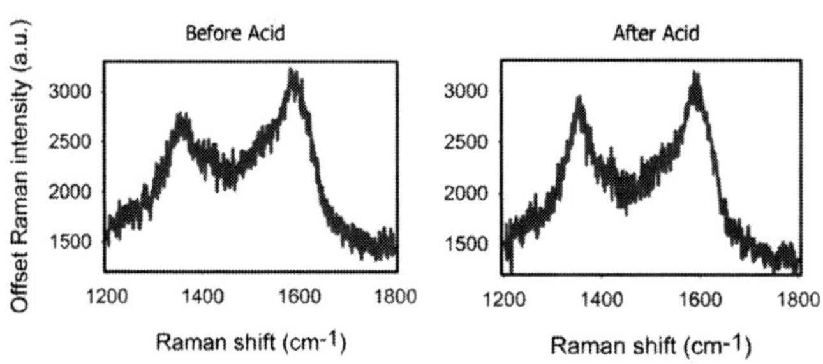

Figure 3: Raman spectra of a single carbon nanotube before and after 1 minute dipping in HNO_3.

An important feature of this experiment is that excitation of the nanotube occurs under TIR conditions. We have calculated the penetration depth of our optical excitation to 100 nm; as such, we are only exciting Raman from the end of the nanotube functionalized AFM tip. By using TIR, we are able to show that the oxidation of the carbon nanotube occurs locally, at the end of the nanotube tip. This is important for future studies, in which the carbon nanotube will be further functionalized using carboxylic acid based coupling chemistry.

CONCLUSIONS

We have introduced an experimental approach for performing wet chemical reactions with spectroscopic detection on a single carbon nanotube, using an AFM coupled to an optical microscope with Raman detection capabilities. Oxidation of a single multiwalled carbon nanotube was demonstrated by dipping the nanotube into an evaporated drop of HNO_3. An increase of the D/G band intensity ratio was observed, which is an indication that oxidation occurred as sp^2 bonds were converted into sp^3 bonds. The rapid oxidation occurred after only one minute, much faster compared to reaction with large amounts of nanotubes. This technique has applications for following reaction dynamics in real time and will prove a useful compliment to established techniques such as chemical force microscopy.

ACKNOWLEDGMENTS

We would like to thank Xidex Corporation, Austin, Texas for the contribution of carbon nanotubes. This work was supported with funding from DARPA (grant no. N66001-09-1-2075) as well as start up funds from the University of Texas at Austin.

REFERENCES

[1] A. Noy, D. V. Vezenov, and C. M. Lieber, Annu. Rev. Mater. Sci. **27**, 381 (1997).
[2] S. S. Wong *et al.*, Nature **394**, 52 (1998).
[3] S. S. Wong *et al.*, J. Am. Chem. Soc. **120**, 8557 (1998).
[4] A. Noy, Surface and Interface Analysis **38**, 1429 (2006).
[5] R. L. Stiles *et al.*, Journal of Physical Chemistry C **112**, 11696 (2008).
[6] A. Javey *et al.*, Nano Letters **4**, 447 (2004).
[7] D. Zhang *et al.*, Nano Letters **6**, 1880 (2006).
[8] M. Endo *et al.*, Carbon **39**, 1287 (2001).
[9] P. M. Ajayan *et al.*, Science **265**, 1212 (1994).
[10] D. Tasis *et al.*, Chemical Reviews **106**, 1105 (2006).
[11] L. Minati *et al.*, Journal of Physical Chemistry C **114**, 11068 (2010).
[12] C. Spudat, C. Meyer, and C. M. Schneider, Physica Status Solidi B-Basic Solid State Physics **245**, 2205 (2008).
[13] Z. Liu *et al.*, Langmuir **16**, 3569 (2000).
[14] D.-H. Jung *et al.*, Carbon **48**, 1070.
[15] M. F. Yu *et al.*, Science **287**, 637 (2000).
[16] R. Garcia, R. V. Martinez, and J. Martinez, Chemical Society Reviews **35**, 29 (2006).

[17] M. S. Dresselhaus *et al.*, Physics Reports **409**, 47 (2005).
[18] M. S. Dresselhaus, ACS Nano **4**, 4344.
[19] S. Costa *et al.*, physica status solidi (b) **246**, 2717 (2009).
[20] S. Chakrabarti, K. Gong, and L. Dai, The Journal of Physical Chemistry C **112**, 8136 (2008).
[21] D. Y. Kim *et al.*, Current Applied Physics **10**, 1046 (2010).

AUTHOR INDEX

SUBJECT INDEX

CPSIA information can be obtained at www.ICGtesting.com
Printed in the USA
BVOW052031111111

275709BV00004B/2/P

9 781605 112954